Photoconductivity

OPTICAL ENGINEERING

Series Editor

---

1. Electron and Ion Microscopy and Microanalysis: Principles and Applications, *by Lawrence E. Murr*
2. Acousto-Optic Signal Processing: Theory and Implementation, *edited by Norman J. Berg and John N. Lee*
3. Electro-Optic and Acousto-Optic Scanning and Deflection, *by Milton Gottlieb, Clive L. M. Ireland, and John Martin Ley*
4. Single-Mode Fiber Optics: Principles and Applications, *by Luc B. Jeunhomme*
5. Pulse Code Formats for Fiber Optical Data Communication: Basic Principles and Applications, *by David J. Morris*
6. Optical Materials: An Introduction to Selection and Application, *by Solomon Musikant*
7. Infrared Methods for Gaseous Measurements: Theory and Practice, *edited by Joda Wormhoudt*
8. Laser Beam Scanning: Opto-Mechanical Devices, Systems, and Data Storage Optics, *edited by Gerald F. Marshall*
9. Opto-Mechanical Systems Design, *by Paul R. Yoder, Jr.*
10. Optical Fiber Splices and Connectors: Theory and Methods, *by Calvin M. Miller with Stephen C. Mettler and Ian A. White*
11. Laser Spectroscopy and Its Applications, *edited by Leon J. Radziemski, Richard W. Solarz, and Jeffrey A. Paisner*
12. Infrared Optoelectronics: Devices and Applications, *by William Nunley and J. Scott Bechtel*
13. Integrated Optical Circuits and Components: Design and Applications, *edited by Lynn D. Hutcheson*
14. Handbook of Molecular Lasers, *edited by Peter K. Cheo*
15. Handbook of Optical Fibers and Cables, *by Hiroshi Murata*
16. Acousto-Optics, *by Adrian Korpel*
17. Procedures in Applied Optics, *by John Strong*

18. Handbook of Solid-State Lasers, *edited by Peter K. Cheo*

19. Optical Computing: Digital and Symbolic, *edited by Raymond Arrathoon*

20. Laser Applications in Physical Chemistry, *edited by D. K. Evans*

21. Laser-Induced Plasmas: Physical, Chemical, and Biological Applications, *edited by Leon J. Radziemski and David A. Cremers*

22. Infrared Technology Fundamentals, *by Irving J. Spiro and Monroe Schlessinger*

23. Single-Mode Fiber Optics: Principles and Applications, Second Edition, Revised and Expanded, *by Luc B. Jeunhomme*

24. Image Analysis Applications, *edited by Rangachar Kasturi and Mohan M. Trivedi*

25. Photoconductivity: Art, Science, and Technology, *by N. V. Joshi*

Other Volumes in Preparation

Principles of Optical Circuit Engineering, *by Mark Mentzer*

Handbook of Lens Design, *by Milton Laikin*

Optical Components, Techniqucs, and Systems in Engineering, *by R. S. Sirohi and M. P. Kothiyal*

Photoconductivity

Art, Science, and Technology

N. V. Joshi
University of Los Andes
Mérida, Venezuela

MARCEL DEKKER, INC. New York and Basel

Library of Congress Cataloging-in-Publication Data

Joshi, N.V.
Photoconductivity : art, science, and technology / N.V. Joshi.
p. cm. -- (Optical engineering : 25)
Includes bibliographical references.
ISBN 0-8247-8321-2 (alk. paper)
1. Photoelectric devices. 2. Photoconductivity. I. Title.
II. Series: Optical engineering (Marcel Dekker, Inc.) ; v.25.
TK8304.J67 1990
621.381'542--dc20 90-2981
CIP

This book is printed on acid-free paper.

MARCEL DEKKER, INC.
270 Madison Avenue, New York, New York 10016

Current printing (last digit):
10 9 8 7 6 5 4 3 2 1

PRINTED IN THE UNITED STATES OF AMERICA

To my wife, Virginia,

and my boys, Joel and Yeewan

About the Series

Optical science, engineering, and technology have grown rapidly in the last decade so that today optical engineering has emerged as an important discipline in its own right. This series is devoted to discussing topics in optical engineering at a level that will be useful to those working in the field or attempting to design systems that are based on optical techniques or that have significant optical subsystems. The philosophy is not to provide detailed monographs on narrow subject areas but to deal with the material at a level that makes it immediately useful to the practicing scientist and engineer. These are not research monographs, although we expect that workers in optical research will find them extremely valuable.

Volumes in this series cover those topics that have been a part of the rapid expansion of optical engineering. The developments that have led to this expansion include the laser and its many commercial and industrial applications, the new optical materials, gradient index optics, electro- and acousto-optics, fiber optics and communications, optical computing and pattern recognition, optical data reading, recording and storage, biochemical instrumentation, industrial robotics, integrated optics, infrared and ultraviolet systems, etc. Since the optical industry is currently one of the major growth industries this list will surely become even more extensive.

Brian J. Thompson
University of Rochester
Rochester, New York

Preface

This book is for readers who are working in interdisciplinary fields of physics, materials science, and optical engineering, dealing with photoconduction and allied phenomena. A reasonable knowledge of quantum mechanics, solid state physics, and group theory is desirable for ease in reading some chapters.

My main objective is to bring to notice several aspects of photoconductivity and photodetection that may have not received due consideration elsewhere and are important not only from theoretical points of view but also from technological points of view. Recent developments provide a newer outlook and help to build a theoretical background for previously reported phenomena.

Over the last few years, developments in the study of quantum well and superlattice structures have dramatically changed the concepts of photosensors, and the new field of modern photodetectors is emerging. It is also my purpose to provide brief, yet useful information about both the basic concepts of photoconductivity and the technological advancements of superlattices, so that readers will be able to grasp the state of the art in this new field.

Photodetection technology is becoming more and more important in military applications, particularly in guided weapons and communication through fiber optics. Infrared developments are based on solid state photonic devices. Further developments in these fields demand a good understanding of the basic principles of photoconductivity processes, and this volume is intended to provide this understanding.

References are listed at the end of each chapter. It is possible to cite hundreds of references on each aspect mentioned in the book;

however, I have selected only a few representative studies, some of which are not frequently cited in the literature. The evaluation of these references has been purely personal, but I have made every effort to include reviews or invited articles whenever possible. By and large, the list contains the references used in writing these chapters, and therefore I acknowledge my debt to these sources. Photoconductivity is such an immense field that I have probably missed relevant and important sources, and if so, I apologize.

An adequate notation is always a vexing problem, particularly in this field. Some notations have been used in different contexts by different authors, and making them coherent could result in ambiguity. In such circumstances, I have differentiated them by capital, lowercase letters and by underlining. When this is not enough, I have used subscripts to explain the meaning explicitly. As far as possible the conventional symbols are maintained throughout. A list of the symbols used here is given at the end of the book.

I thank Dr. Juan Aponte of the Instituto Venezolano de Investigaciones Cientificas and Professor L. Laude, Université de l'État, Mons, Belgium, for reading the manuscript carefully and making valuable suggestions. I sincerely appreciate the discussion with Dr. H. K. Henisch of some sections of the book. His critical comments were more than helpful. I also thank the members of the editorial department of Marcel Dekker, Inc., for their excellent collaboration during the production of the book.

In the last several years, my projects on Photoconductivity were funded by two organizations: Consejo Nacional de Investigaciones Cientificas y Technològicas (CONICIT), Venezuela, and Consejo de Desarrollo Cientifico, Humanistico y Tecnològico (CDCHT) of University of Los Andes. This book is the output of the projects and, therefore, I thank those organizations.

Writing a book is always a long-term project and needs constant encouragement, which I received from my wife, Virginia. Moreover, by improving the language, she helped to convert my notes into manuscript.

I hope that readers will enjoy this book. Needless to say, comments and criticisms are always welcome.

N. V. Joshi

Contents

Photoconductivity

1
Introduction

1.1 INTRODUCTION

Photoconductivity is an important property of semiconductors by means of which the bulk conductivity of the sample changes due to incident radiation. When the conductivity decreases, then "negative photoconductivity" is the term used. There are a few experimental situations where such an effect is observed, but in general the conductivity increases when the radiation is incident on the sample. This property has been used advantageously in several branches of science and technology.

Photoconductivity is not an elementary process in solids. Photoconduction, as the name suggests, includes the generation and recombination of charge carriers and their transport to the electrodes. Obviously, the thermal and hot carrier relaxation process, charge carrier statistics, effects of electrodes, and several mechanisms of recombination are involved in photoconduction. Above all, every mechanism mentioned here is a complicated one, and therefore photoconductivity in general is a very complex process. For interpretation purposes, complementary information from other experimental techniques such as photoluminescence, optical absorption, and electroreflectance is always needed.

In spite of the complexity of the photoconductance process, it provides useful and valuable information about physical properties of materials and offers applications in photodetection and radiation measurements. Recent advances in thin-film technology, particularly in excellent quality crystal growth (doped and undoped) by metal organo chemical vapour deposition (MOCVD) and molecular beam epitaxy (MBE) gave a new dimension (in fact, a reduced dimension) to the photodetectors, and now quantum well and superlattice photodetectors are a reality. We will see in Chapter 6 some of the details of modern photodetectors. Here we want to note that all these features make photoconductivity a fast-growing field.

Detection of weak photosignals needs a careful selection of several parameters. It is necessary to know the change in the resistance of the sample caused by the incident radiation, the impedance of the measuring circuit, the response time, the capabilities and limitations of the photodetector, and sources of noise with their magnitudes. The combined knowledge helps in selecting the proper instruments. The handling of such information and its use for low-level radiation detection requires a conceptual foundation, and therefore they are certainly as much an art as a science.

In modern life, many applications demand a time domain of a short interval (of the order of picoseconds). This demands that photodetectors have ultrafast response. To achieve this, the distance between two contacts must be small (on the order of a micrometer), a technological requirement. In short, photoconductivity is science, art, and technology combined.

Historically, the first photoconductivity effect was recorded in 1873 by W. Smith [1], who observed that the resistivity of selenium was decreased by radiation shining on it. According to the literature, this is the very first experimental detection of photoconductivity (in fact, it was a combination of photoconductivity and the photovoltaic effect).

Understanding the origin of the observed photoconductivity was very difficult at that time. At the end of the nineteenth century, it was known experimentally that when radiation (visible or ultraviolet) is incident on a metal surface, electrons are ejected from the metal. This observation was not so unexpected, but what was surprising was that the kinetic energy of the ejected electrons is proportional to the frequency of the radiation and independent of its intensity. This phenomenon could not be understood on a classical basis.

In 1905, Einstein explained this puzzling observation. A beam of monochromatic light consists of packages of energy of magnitude $h\upsilon$ where h is Planck's constant and υ is the frequency of the radiation. During the interaction process this quantum of energy is completely transferred to the electron. Thus, the electron acquires energy $h\upsilon$. The kinetic energy of the electron, therefore, is equal to $h\upsilon$ minus the work function of the metal.

Considering the equivalence of mass and energy, the momentum of each quantum (or photon) can also be calculated as $h\upsilon/c$ or h/λ.

Einstein's explanation of the photoelectric effect thus opened the way to interpreting the interaction of radiation with matter, particularly optical absorption, photoconductivity, the photovoltaic effect, and other related phenomena. This effect was considered an interesting property of selenium. Soon, more and more materials having this property were discovered, and today we know that there exist more than 1000 semiconducting compounds that have a reasonably good photoresponse. In fact, all nonmetals have a certain degree of photoresponse, even though in many cases it is difficult to detect.

From about 1920 to 1940 the photovoltaic effect was also investigated. In a short time, a close relationship between rectifier and photovoltaic cell was realized, and around the beginning of 1930 photocells of copper and selenium became commercially available. Immediately the importance of photocells became clear, particularly for defense; and the intensive and systematic investigation of radiation detection was accelerated. During World War II, attention was focused on infrared detectors, but that work was classified. After the end of the war, somewhere around 1947, this work was declassified. The major development in this period was carried out on lead compounds such as PbS, PbSe, and PbTe.

Following World War II, a rapid development of several branches of solid state physics, among them photoconductivity, is observed.

The basic principle involved in photoconductivity is often stated in a very simple way. When photons of energy greater than that of the band gap of the semiconductor are incident upon a photoconductive material, electrons and holes are created in the conduction and valence bands, respectively, increasing the conductivity of the sample. But this statement is only partially true. In a doped semiconductor, for example, the photon of slightly less energy than that of the band gap is absorbed by the impurity atom and a free electron is created in the conduction band [2,3]. In this case, photoresponse starts from the low-energy side of the band gap and photoconduction occurs due to excitation near the band edge. This is a general case and not an exception.

It is also possible to observe photoconductivity when the energy of the incident photon is much smaller than that of the band gap. When the energy of the incident photons matches the ionization energy of the impurity atoms, they are ionized, creating extra electrons in the conduction band, and hence an increase in conductivity is observed. This phenomenon is called extrinsic photoconductivity and lies in the far-infrared region. The detectors based on this principle are described by Levinstein [4]. Several interesting aspects of extrinsic photoconductivity, particularly its relations with optical phonons, have been discussed by Stradling [5].

Absorption of radiation—a basic process of photoconduction—is not limited to the processes just discussed. There exist several complementary mechanisms that create free charge carriers. A typical example is the recently investigated diluted magnetic semiconductors (ternary alloys of group II—VI compounds with transition metals such as $Cd_{1-x}Mn_xTe$, $Cd_{1-x}Mn_xSe$, $Cd_{1-x}Mn_xS$, etc.), which reveal an interesting property [6]. Some 3d levels and excited states of transition metal ions are pinned with respect to the band edge of the alloy. The precise position of the level depends on the alloy and the concentration of the transition metal. If 3d levels are pinned near the edge of the band gap, then the photoconductivity originating from this state merges with the impurity photoconductivity. Very often, 3d levels are pinned deep in the valence band. When the energy of the incident radiation coincides with the energy difference between one of the 3d levels and the bottom of the conduction band, then the photons are absorbed, resulting in an increase in conductivity. Let us call this type of photoconductivity "inner-level excitation photoconductivity." It has already been reported in $Cd_{1-x}Mn_xSe$ [7] and a few other diluted magnetic semiconductors [8]. In short, photoconductivity is due to the absorption of photons (either by an intrinsic process or by impurities with or without phonons), leading to the creation of free charge particles in the conduction band and/or in the valence band. Photoabsorption and hence photoconduction take place by one of the following mechanisms.

1. Band-to-band transitions
2. Impurity levels to band edge transitions
3. Ionization of donors
4. Deep level (located in the valence band) to conduction band transitions

The mechanism of the absorption essentially depends on the details of the band structure, and hence separate attention is given to this in Chapter 4 for conventional and nonconventional materials.

The concept of band-gap (or band-to-band) transitions in the optical absorption process and the creation of free charge carriers can be extended to amorphous semiconductors, with the difference that the density of states in amorphous materials is different from that of crystalline materials. Obviously, the details of the photoconducting properties for amorphous and crystalline materials may vary but they largely show similar tendencies.

The above discussion does not hold for generation and transport of the charge carriers in organic compounds. These materials have poor and very often slow photoresponse and therefore are not adequate for photodetection purposes. However, they have other uses such as in copying machines and electrophotography where the latent images are produced on large-area photoconductor films. In organic photoconductors, charge generation takes place through transitions in molecular orbitals. This is a completely different situation from that of inorganic crystalline and amorphous materials.

Moreover, recombination and transport mechanisms of the charge carriers are also entirely different in organic photoconductors. For these reasons, organic photoconductors, even though they are useful from the technological point of view, will not be discussed in this book. Attention is mainly focused on the materials that are potentially important for photodetection and materials technology.

Deep-level excitation photoconductivity has unexplored potential applications. For technological use, the photoresponse of semiconductors is increased and is shifted to the lower-energy side by introducing adequate impurities, either acceptors or donors, but in diluted magnetic semiconductors the photoresponse can be shifted to the higher-energy side by incorporating transition metal ions. This is a very useful property for the detection of ultraviolet radiation with solid state devices, for which the search for good quality materials is constantly being carried out. A high photoresponse on the higher-energy side than that of the band gap has been already reported in $Cd_{1-x}Mn_xS$ [9]. For $x = 0.1$, the band gap at 300 K is 1.8 eV, while a high photoresponse has been observed on the higher-energy side of the band gap [1.9 eV]. This aspect will be discussed later.

In addition to the understanding of the basic processes involved in photoconductivity, constant efforts are being made to search for new materials. First, alkali halides were investigated. Then, because of industrial applications, interest was focused on elemental (germanium and silicon) [10] and compound semiconductors such as the family of group II—VI (CdS, CdSe, ZnS, ZnSe, etc.) [11] and III—V compounds (GaAs, InP, etc.) [12] and, recently, ternary and multinary compounds [13]. It is worth mentioning that some of the ternary compounds show a really high photoresponse, ensuring future applications, some of these will be discussed in Chapter 4. The rapid progress in concepts, materials, and technology is indicated by an exceptionally large number of monographs and review articles published since 1960 [14—34].

Photodetector technology is a very complex issue. The selection of the proper material is a key parameter and depends upon many factors. Complete information on electronic and optical properties such as spectral dependence of absorption coefficient and reflectivity should be available. The parameters involved in charge carrier statistics (lifetimes of majority and minority carriers, position of Fermi level in doped and undoped materials, etc.) and their dependence on the type of impurity and temperature should also be known. If the interface is involved, then precise knowledge of lattice parameters and their variation with temperature are important factors to achieve heterojunction with a minimum number of defects. Above all, the technology of high quality crystal growth is a necessary requirement. This includes material purification, control over the growth process, doping and annealing techniques, and a set of methods of characterization and evaluation of the quality of the crystal. No wonder that to describe some aspects of a given photodetector (e.g., mercury cadmium telluride [35] or indium antimonide [36]) a separate review article is required. This illustrates the overall complexity of the physics and technology of photoconductors. In spite of these complications, the technology of photodetection is very precise and mature enough for several purposes (see Chapter 6).

One of the reasons for the rapid development of this field is its applications in industry and defense. Now it is well known that highly sensitive detectors in the infrared region are generally photoconducting types. The use of pn photovoltaic devices is also very common in science and technology because of their small physical dimensions, high responsivity, and high speed. Phototransistors are natural offsprings of the pn junction device family.

Recently, a large amount of work has been carried out on optical communication and optical computers [37]. For these purposes (and also for many others), integrated electrooptical circuitry [38] is being developed. One of the important features of this is the integration of monolithic ultrahigh-speed photonic devices [39] (laser diodes and high-speed photodiodes such as quantum well and superlattice detectors [40,41]). According to recent publications, such capability exists. I am confident that in a short period such devices will be available commercially.

With these introductory comments I want to emphasize that the art of radiation detection requires a knowledge of new materials, new techniques, new devices, and also new approaches. After 115 years of investigations on photoconductivity, the basic mechanisms of charge generation and their transport are reasonably well understood. However, there are several problems such as transient photoconductivity and contact and surface effects on photoconductivity that are not satisfactorily understood. Even though partial understanding is achieved, further progress in this direction is needed for improving device performance.

In spite of 100 years of progress, a few terms are not properly defined. Different authors use these terms in a different sense, leading to ambiguity. It is therefore necessary to define the terms that will be used constantly in this book.

1.2. ANALYSIS OF THE TERMINOLOGY

in photoconductivity sometimes common terminology is misused. This is because many of the terms refer directly or indirectly to the parameters of the kinetics of the charge carriers, and because they are interrelated in a complicated manner (see Chapter 3), different authors use different terminology according to the assumptions involved in a specific context. Frequently, these assumptions are not explicitly mentioned.

Naturally, in a simplification process, minute differences are overlooked. A typical example is the term "time constants." There are a few such as lifetime, response time, dielectric time, and pair lifetime that appear in the literature. The first three terms are likely to be confused, as they have the same meaning in some circumstances. Unfortunately, the notation used to describe them is not well defined, and different authors have used different terminology. To avoid misunderstanding, in this volume the terms are explained more explicitly and the range of their applicability for photoconductivity and related phenomenon is discussed. Complete understanding of some of these terms requires more knowledge about the kinetics of traps, which will be discussed in Chapter 3.

1.2.1 Absorption Coefficient

The very first stage in the operating mechanism of photoconductivity is the absorption of incident radiation quanta of appropriate energy with the generation of free charge carriers. This is the basic principle of converting radiation into electrical energy. There are several processes in which photons are absorbed in the matter. Independent of the process of absorption, the intensity of radiation (photon flux) decreases with distance inside the material according to the exponential relation known as Lambert's law,

$$\underline{I}(x) = \underline{I}_0[1 - R(\lambda)]\exp(-\alpha x) \tag{1.1}$$

where $R(\lambda)$ is the reflectivity at wavelength λ; $\underline{I}_0$ and $\underline{I}$ are the intensities or photon fluxes at the surface of the material and at a distance x within the matter, respectively; and α is the absorption coefficient, measured in cm^{-1}.

1.2.2. Dark Current

Dark current, I_d, is the amount of current that flows through the semiconductor or device when no radiation is incident on it. It changes with operating temperature and applied voltage, and therefore these parameters should be always mentioned. Dark current is not a constant background current but also has fluctuations or noise. The average d.c. value of the current is generally mentioned as dark current. Unfortunately, the level of fluctuation, which is an important parameter, is rarely mentioned.

1.2.3. Signal Response

Signal response, R(λ), is the difference between the dark current and the output of the photosensor when it is illuminated with radiation having energy higher than the band gap. It is measured in volts or amperes. The magnitude of the signal response depends on the device, the intensity of the radiation, and the applied voltage.

1.2.4. Responsivity

The term responsivity is used to describe the response of the photodetector, normalized to incident power and measured in either volts per watt or in amperes per watt. It is the ratio of the output voltage or current (according to the circuit of measurement) of the photodetector to the input radiant power in watts, photons, or lumens. Given the wavelength of incident radiation, the number of photons can be obtained from the measurement of radiation in W/cm^2. Currently, the "lumen" unit is not used.

The responsivity of a given detector depends on the following factors.

1. Size of the active area of the detector. With a large area, more light is collected and converted into electrical output signal.
2. Internal gain of the device. This is defined as the number of free charge carriers at the output of the device to the number of photogenerated charge carriers. Modern photodetectors such as the avalanche photodiode have a high internal gain and hence response is higher. A typical phototransistor has 500 gain while a photodarlington (a pair of transistors packed together so that the output of one phototransistor is the input to the other) could have a gain as high as 10,000.
3. Intensity of the incident radiation. Some devices have a linear response in a limited range of intensities. At high intensity, the response is saturated. The range of linear response and limit of saturation vary from one material to another and from device to device. It is therefore necessary to mention the incident power for which the responsivity is measured.
4. Electrical circuit employed in measurement. Biasing circuitry and load of the detector also affect the responsivity of a given detector.
5. Assumption that the contacts are ohmic. (Here ohmic means contacts that follow Ohm's law; in many devices, Schottky type contacts are used, and these are excluded.) If the contacts are ohmic, the responsivity increases linearly with the applied electric field. In practice, the upper limit is set up by Joule heating resulting from the applied bias. This consideration is necessary for evaluating the responsivity [36].

Sometimes the responsivity is given as a plot of resistance versus illuminating power rather than as a number. This is very useful, particularly when the variation of resistance with respect to incident power has a slight deviation from linearity. Of course, the slope depends on the operating temperature and sometimes on the incident power. A typical curve for commercially available detectors is given in Fig. 1.1. The user knows directly from this plot the useful intensity variation and linearity. Generally, for a wide range of illumination, the resistance is linear, so the slope of the line can be used as an indicative figure of the responsivity. This number is defined as

$$\gamma_{10}^{100} = \frac{\log(R_{10}/R_{100})}{\log(10/100)} \tag{1.2}$$

It is assumed that in this range the slope of the line is not altered.

It is very important to realize that a change in the resistance is not always instantaneous and it takes some time to attain the resistance specified by the manufacturer. Although it is not customary to mention this time, it is important to do so. The terms "signal response" and "responsivity" should not be misused. The signal response can be improved by aligning the optical components or by

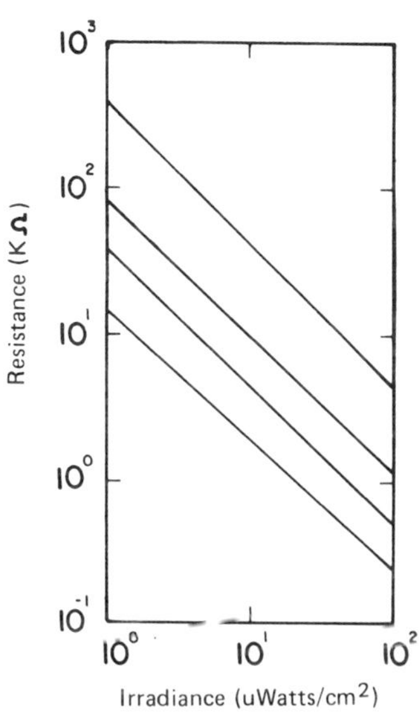

FIGURE 1.1 Variation of resistance of a photocell with intensity of illumination.

introducing an additional optical element such as a lens, or mirror. However, responsivity of the sensor is a property of the sensor and will not be altered by external experimental variations.

1.2.5. Spectral Response

The relative sensitivity (responsivity) of a photoconductive or photovoltaic cell depends on the wavelength of the incident radiation. A plot of the relative response as a function of wavelength is called a spectral response curve. Typical curves for a CdS/CdSe photoconductive cell are given in Fig. 1.2. Figure 1.3 shows spectral response curves for some conventionally used photodiodes. Responsivity is given in either absolute units (A/W) or in relative units, where the peak response is considered as 100% response. The position of the peak response approximately corresponds to the band-gap value of the semiconductor and depends on the photoconductive material. On the longer-wavelength side, the photoresponse goes down very fast, because the photons do not have enough energy to create electron–hole pairs. It is frequently observed that the response goes down on the shorter-wavelength side also even though the absorption coefficient is high in this spectral region. This is because the photons of these energies are absorbed at or near the surface of the semiconductor, where the recombination

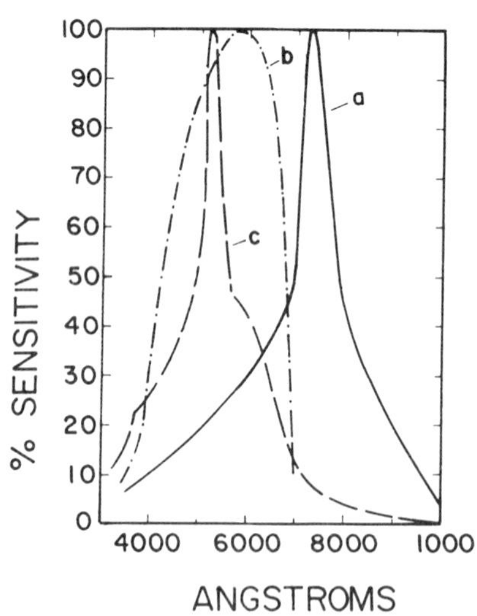

FIGURE 1.2 Relative spectral response of commonly used photoconductive detectors. (a) CdSe; (b) CdS; (c) GaAsP.

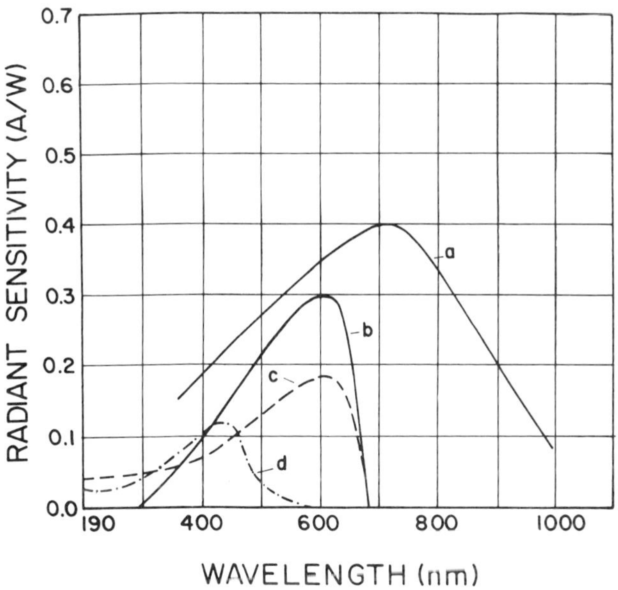

FIGURE 1.3 Spectral response of a few photodiodes, which cover the spectral range from ultraviolet to near-infrared. (a) Silicon photodiode; (b) GaAsP photodiode; (c) GaAsP, Schottky type; (d) GaP ultraviolet photodiode. (Spectral response and sensitivity vary with the alloy composition, technology and design. A substantial change in both spectral response and sensitivity is expected, according to the manufacturer. [Reprinted with permission from Hamamatsu Photonics, K.K., Japan.]

velocity is higher than in the bulk. However, the spectral response can be extended to the high-energy side by selecting a proper (or split) valence band structure of the photoconducting material. This has been reported in ternary semiconducting compounds [42,43].

1.2.6. Angular Response

Responsivity is stated for incident radiation that is normal to the plane of the sensitive surface. When the angle of incidence varies from the normal, the responsivity decreases according to the cosine of the angle, and the flux of the radiation varies in the same manner. Responsivity also varies with the wavelength. This is because the reflectivity of the active area (or of the thin layer of coating evaporated in the fabrication process) varies with the incidence angle. Even though the deviation of angular response with respect to the normal is small (10%), a separate calibration is needed or it is recommended that these data be obtained from the manufacturer.

1.2.7. Signal Current

The signal current, I_s, of a photodetector is given by

$$I_s = R(\lambda)\,\phi_{e\lambda} d\lambda \tag{1.3}$$

where $\phi_{e\lambda}$ is the incident flux at a given wavelength λ. Sometimes the flux is given for a range of wavelengths; in that case,

$$\phi_e = \int_{\lambda_1}^{\lambda_2} \phi_{e\lambda} d\lambda \tag{1.4}$$

1.2.8. Quantum Efficiency

Quantum efficiency (η) is defined at a particular wavelength and is the ratio of the number of photoelectrons created per second to the number of incident photons per second. It is necessary to be careful about using this definition because some authors use the number of absorbed photons rather than incident photons. From the point of view of the users, "incident photons" is more accurate since in the laboratory the flux of incident photons is controlled. If other parameters of the detector are known, the number of absorbed photons can be calculated. It is worth mentioning that this term refers to the primary process in the device, that is, the interaction of photons with the sensor. Internal (e.g., avalanching photodiode) or external amplification, even though it is carried out on the same chip as in monolithically integrated devices, should be excluded. Naturally, the maximum value of quantum efficiency should be 1.

The number of incident photons per unit time per unit area is expressed by

$$P = N_{pho} h\nu \tag{1.5a}$$

$$N_{pho} = \frac{\lambda}{hc} P \tag{1.5b}$$

where P is the radiant flux density in watts per centimeter square and N_{pho} is the number of photons $cm^2 \cdot sec$. Using this equation, quantum efficiency can be given by [15]

$$\eta = \frac{hc}{e\lambda} R(\lambda) \tag{1.6}$$

where $R(\lambda)$ is responsivity (A/W), e is the charge of the electron, and the other symbols have their usual meaning. Here it is assumed that all the incident radiation is absorbed in the given photosensor. This is a very useful formula for calculating quantum efficiency, because the spectral response $R(\lambda)$ is generally available.

1.2.9. Cutoff Wavelength

This value represents the long-wavelength limit of spectral response. Generally, it is the wavelength for which the response is 10% of the value at the maximum response wavelength. This concept should not be taken too seriously because it does not explain how far the response extends and where it really becomes nearly zero. Even though the responsivity is less than 10% of the peak response value, the corresponding region can be useful for photodetection purposes. Moreover, cutoff wavelength does not inform on the responsivity at that wavelength but on the relative responsivity compared to the peak value and hence has no relation to the capability of photodetection.

1.2.10. Light History Effect

Normally the light history effect is observed in photoconductive cells, particularly in wide-band-gap materials ($E_g > 1.6$ eV). This is also known as fatigue or the hysteresis effect. Occasionally, the term "light memory effect" is also used, but recently this term has come to be used to refer to a slow component of the photoconductivity decay curve or to persistent photoconductivity, and hence we will not be using it for light history effect.

Both the response time and sensitivity of photoconductive detectors vary with the "previous" conditions of radiation. The magnitude of the effect depends on the present light level, the difference between the present and previous light intensities, and the duration of the radiation. Even though this effect is a hindrance in the use of photoconductive cells in light-measurement applications, it is not a significant problem in the usual applications at high levels of radiation and where on-off switching is important. However, this knowledge is necessary to evaluate the term "responsivity." It is not usual, but it is useful to mention the time elapsed between removal of incident illumination (in watts or lumens) and the measurement of responsivity.

1.2.11. Response Time

The response time of a photoconductor or photoconducting device is a measure of the time required for the photocurrent to increase (decrease) after radiation of proper energy (enough to excite electrons and make them free) is turned on (off). Let us assume that photocurrent increase (decreases) in an exponential manner with a single time constant. Then the rise and fall of the photocurrent can be expressed as

$$I(t) = I_0(1 - e^{-t/\tau_{res}}) \quad \text{(rise)} \tag{1.7a}$$

or

$$I(t) = I_0 e^{-t/\tau_{res}} \quad \text{(decay)} \tag{1.7b}$$

where $I(t)$ and I_0 represent the photocurrent as a function of time and its maximum value, respectively, and τ_{res} is a time constant. For $t = \tau_{res}$, the current reaches 63% of its maximum value in the rise curve (37% in the decay curve). It is therefore defined as the time required to attain 63% of I_0 (see Figs. 1.4 and 1.5).

It can be seen that this time constant is a very important parameter not only because it gives information about the speed of the response but also because it is related to the recombination process of the excess charge carriers. It is worth pointing out here that it varies with the intensity of the incident radiation. However, it is assumed in Eqs. (1.7) that the excitation is sufficient low so that τ_{res} is not altered during the illumination process.

The response of photoconductors often deviates from exponentiality, for several reasons; for example, the presence of defect states perturbs or alters the recombination process. If there exists a photomemory effect or optically active traps, then it takes a long time to attain a steady state value after the radiation is turned off. In such cases, experimentally observed rise and decay forms are as shown in Fig. 1.6 (not exponential). Obviously, here the response time loses its meaning (it is not possible to know how much time is required to have saturated photoresponse, and even the form of the curve cannot be guessed from the response time constant). Therefore, in such circumstances, it is not desirable to use this parameter for describing the speed of the sensor.

1.2.12. Rise/Fall Time

Rise or fall time is the time required to increase the photocurrent from 10% to 90% of its final value, or to decrease it from 90% to 10%,

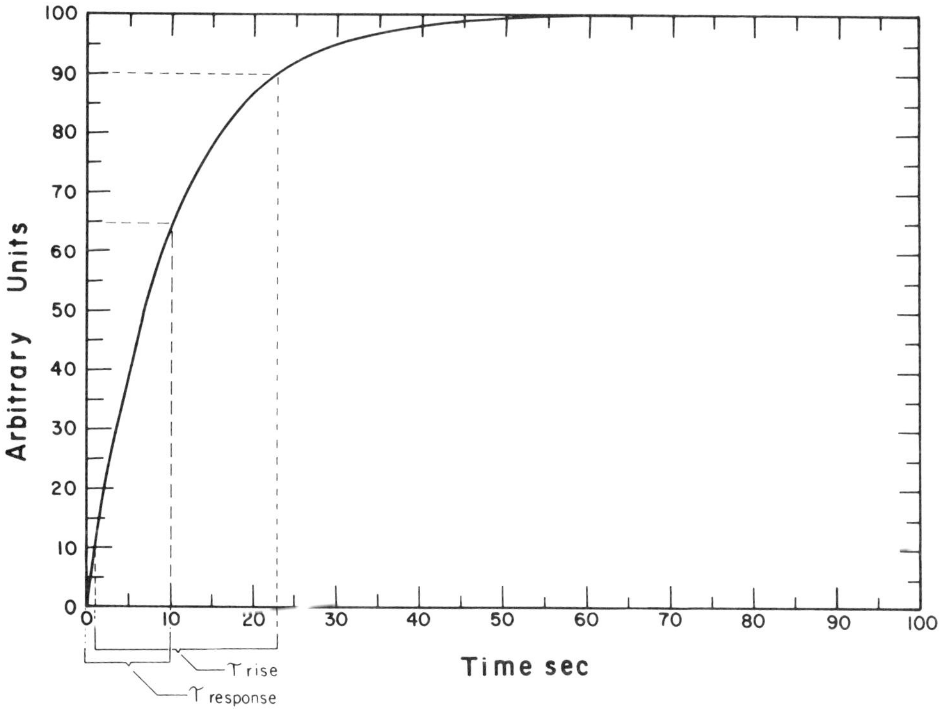

FIGURE 1.4 Exponential rise of photocurrent. The difference between $\tau_{response}$ τ_{rise} is indicated.

respectively. This parameter plays an important role in determining modulation frequency in experimental measurements. The difference between response time and the rise time constant can be understood with the help of Fig. 1.4.

Both time constants depend on the incident light level, load resistance, working temperature, and previous environments of the photoconductors. Generally, photoresponse is faster when the intensity of the radiation is higher. It is also a commonly observed fact that photosensors kept in a dark environment for a long time show slower response than those kept at brighter levels. The precise behavior depends on the statistics of the excess of charge carriers and the density and distribution of optically active traps. Pertinent information about both parameters is difficult to acquire, and hence it is very difficult to estimate rise and fall time, even qualitatively.

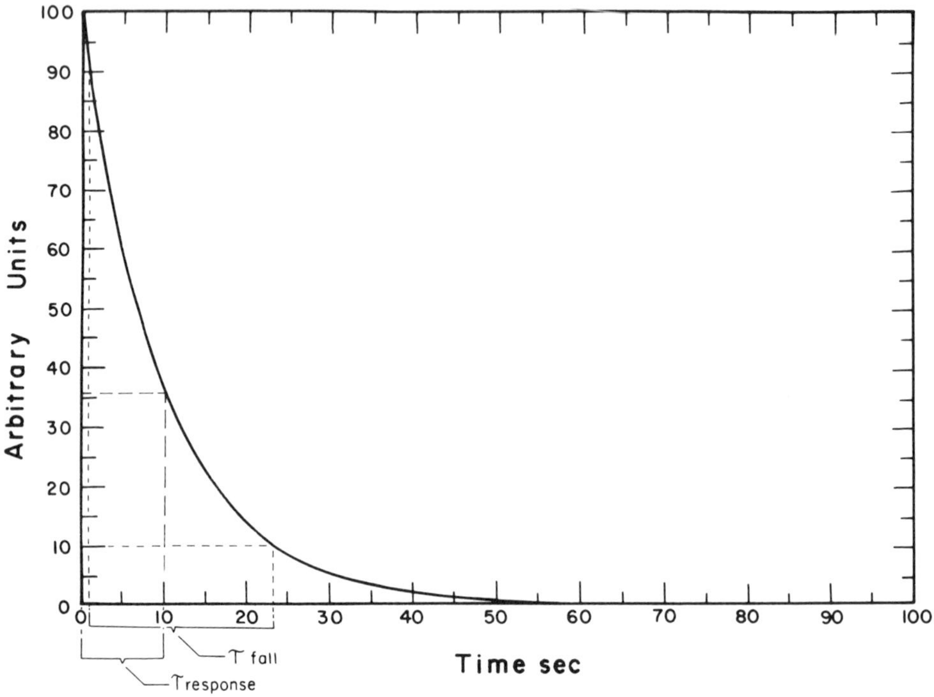

FIGURE 1.5 Exponential fall of photocurrent when radiation is turned off.

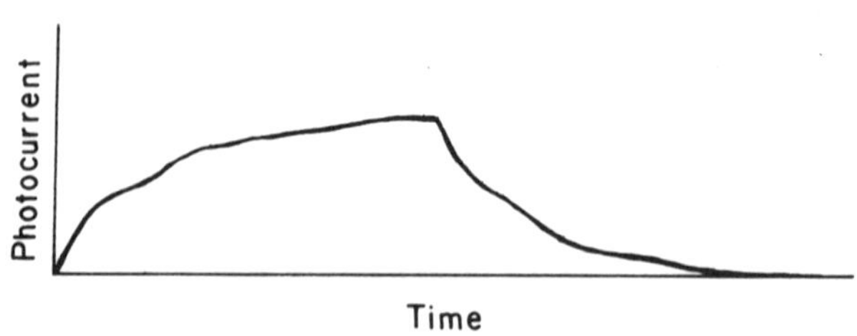

FIGURE 1.6 Nonexponential, but usually observed, time dependence of photocurrent.

1.2.13. Relaxation Time Constant

A basic approach to understanding the transport of the photoexcited charge carriers is to make what is known as a relaxation time approach. When the sample is irradiated, its resistance does not decrease instantly but takes a certain time to attain its original value. This relaxation time can be obtained by considering the rate equation of the photogenerated charge carriers. It is based on the rates of two competing processes: the rate at which the charge carriers are created in the conduction and valence bands and the rate at which they recombine. The net effect, in the absence of active traps, can be expressed by

$$\frac{d\underline{n}}{dt} = \eta\alpha(\lambda)\underline{I}_0 - \frac{\underline{n}}{\tau_{rel}} \tag{1.8}$$

where τ_{rel} is the relaxation time constant, which is inversely proportional to the rate of recombination. The solution of this equation is similar to that of Eq. (1.7) and shows an exponential rise or fall. The value of the response time τ_{rel} can be obtained very easily.

Even though Eq. (1.8) is correct as it stands, it may give a misleading impression of the process. In fact, it is necessary to consider $\underline{n}$ as a function of both time and space. Electrons are captured in traps or in recombination centers and constitute a space charge, which modifies the field. Moreover, here we have to assume that τ_{rel} is a constant, but it could be a function of $\underline{n}$. Equation (1.8) is oversimplified to illustrate the meaning of the term τ_{rel}.

If and when the kinetics of the charge carriers follow Eq. (1.8), the rise and decay curves will have an exponential form with response time equal to τ_{rel}.

The relaxation lifetime approach is by no means always a correct approach. If the photoexcited charge carriers decay by different mechanisms, then they have different decay rates, in which case there can be no unique τ_{rel} that can apply to all possible mechanisms. Since decay rates are different for different groups of electrons, there is no reason to suppose that photocurrent will always decay exponentially with time. This aspect will be reconsidered in detail in Chapter 3. Great care must be taken to correlate response time with relaxation time, particularly when the observed relaxation curve deviates from being exponential.

1.2.14. Collision Time Constant

The collision time constant τ_{coll} is essentially different from the previous two constants. It is not the relaxation time, and there is no decay process involved in it. Because of this fundamental difference, τ is not the proper symbol to use; however, in order to be consistent with the literature we follow the same nomenclature. Usually the lifetime is defined as the reciprocal of the coefficient of the nonequilibrium carrier density in the rate equation (1.8). This can be understood if an alternative approach is adopted to explain charge carrier transport in semiconductors, where it is the mean free time of the electrons in the conduction band (or holes in the valence band), the time between two successive collisions. The mean free time can be obtained by considering either the average motion of one electron over a long period of time or the average motions of many electrons at a particular time. The number of electrons involved in the averaging process is very large, and therefore time average and ensemble average necessarily lead to the same result [44].

Let us consider the ensemble average of n electrons that have suffered a collision at time t_0 and assume that out of them only n electrons have not undergone a subsequent collision during the time interval $t - t_0$. The probability of a collision for an electron (hole) in the time interval dt is given by $(n\,dt)/\tau_{coll}$. Hence, the number of scattered electrons in time interval $t - t_0$ is

$$dn = -\frac{n\,dt}{\tau_{coll}} \tag{1.9}$$

This rate equation applies for free (noninteracting) electrons. This means that the concentration of electrons is low, $\leq 10^{18}/\text{cm}^3$. The solution of Eq. (1.9) is given by

$$n = n_0 e^{-(t-t_0)/\tau_{coll}} \tag{1.10}$$

The average time between collisions can be calculated easily:

$$\langle t - t_0 \rangle = \frac{1}{n_0} \int_{t_0}^{\infty} (t - t_0) \frac{n}{\tau_{coll}}\, dt \tag{1.11}$$

Substituting Eq. (1.10) into (1.11), we get

$$\langle t - t_0 \rangle = \frac{1}{n_0} \int_{t_0}^{\infty} (t - t_0) \frac{n_0}{\tau_{coll}}\, e^{-(t-t_0)/\tau_{coll}}\, dt \tag{1.12}$$

Letting $t - t_0 = x$, we obtain

$$\langle t - t_0 \rangle = \int_0^{\infty} \frac{x}{\tau_{coll}} e^{-x/\tau_{coll}} dx \qquad (1.13)$$

Since this integral is a gamma function in Euler's form,

$$\langle t - t_0 \rangle = \tau_{coll} \Gamma_2 = \tau_{coll} \qquad (1.14)$$

Thus, when the decay process strictly follows Eq. (1.9), it represents the average mean free time of the electrons in the conduction band, that is, the average free time between two successive collisions. The solution of this equation is similar to that of Eq. (1.7) (after adjusting the constants) and the relaxation curves have exponential form. In this case only, τ_{res} represents the average mean free time and the time required to dissipate excess momentum.

In dealing with the term "mobility," the time constant τ_{coll} is used. This time constant could be a few orders of magnitude different than the relaxation time constant. The difference is notable in impurity photoconductivity, a process in which recombination is dominated through defect states. In these cases, the relation between τ_{rel} and τ_{coll} has been reported [14].

The difference between these two constants can be understood with the help of a simple example. In modern photodetectors, efforts are made to increase the mobility (i.e., τ_{coll}) and reduce the response time, and this can be achieved.

1.2.15. Lifetime

The concept of lifetime and its relation to other time constants has been developed by several investigators and thoroughly discussed in several textbooks [14,16] and articles. The term "lifetime" has been defined in different ways by different investigators. Ryvkin [14] explained clearly the difference between relaxation time and the lifetime of charge carriers. Unfortunately, the same symbol used in the literature for both time constants, which creates confusion. The essential difference between the two constants will be clear after we analyze the kinetics of photogenerated charge carriers in Chapter 3.

Bube [11] has defined four types of lifetime constants:

Free lifetime is the time spent by the free electron (hole) in the conduction (valence) band. This is the time that individual electrons (or holes) participate in the conduction process.

Excited lifetime is the time between the act of excitation and the act of recombination (whether radiative or nonradiative). The time spent by the electron in the trap is also included.

Pair lifetime is the free lifetime of an electron—hole pair. This concept is generally used when the numbers of photoexcited electrons and holes are equal ($n = p$). This condition is fulfilled only when the concentration of traps is very low. For trap-dominated photoconductors the use of the term "pair lifetime" is not justified at all.

Minority carrier lifetime is the free lifetime of the minority carriers, and sometimes it is equal to pair lifetime. Its relation to lifetime in the presence of traps can be estimated using the Shockley-Read theory.

1.2.16. Transit Time

Transit time is defined as the time required for the charge carriers to reach the electrode after they are generated in the semiconductor. This depends on their mobility, the distance between electrodes, and the applied voltage. Transient time is expressed as

$$\tau_{transit} = d^2/\mu V \tag{1.15}$$

where μ is the charge mobility, V is the applied voltage, and d is the distance between electrodes.

In obtaining Eq. (1.15), it is assumed that there is no diffusion of the charge carriers and that the field is homogeneous between the two electrodes. In fact, these assumptions are not realistic. The second assumption is violated in several experimental situations, particularly in modern photodetectors, and care should be taken to include the nonuniformity of the field. We will see that this is a key parameter in high-speed photodetectors, where attempts have been and are being made to reduce the transit time. Obviously, this demands good quality material (high mobility) and technology to make electrical contacts at a distance of a few micrometers. The importance of transit time and its applications will be clarified in Chapter 6.

1.2.17. Dielectric Relaxation Time Constant

In fact, the dielectric relaxation time constant is not directly related to the time constants involved in the photoconductivity process. However, in experimental measurements it limits the photoconductivity response time, and therefore its understanding is necessary.

In a homogeneous semiconductor, space charge neutrality is observed; that is, if an unbalanced excess of charge density is created for any reason, then it will be eliminated exponentially, with a time constant known as the dielectric relaxation time constant. Therefore, knowledge of this constant is pertinent to photoconductivity studies.

Let us consider an n-type material and further assume that the electron concentration is uniform at equilibrium (n_0) throughout the material. This assumption is generally valid in photoconducting devices, but care should be taken when considering modern devices where concentration varies in micrometer range. The photoinjected carrier density is and is small as compared to its equilibrium value n_0. It means that the space charge concentration caused by the radiation is $e\underline{n}$, and, according to Gauss's law, generates an electric field E.

According to the continuity equation [45],

$$\nabla \cdot I_n = -e \frac{\partial \underline{n}}{\partial t} \tag{1.16}$$

where I_n is the electric current density and is given by

$$I_n = e\mu_n \underline{n} E + eD_n \nabla \underline{n} \tag{1.17}$$

Assuming that the excess of charge carriers generated is uniform over the sample (this is a crude assumption and not really satisfied), we can ignore the contribution from the diffusion current and write

$$I_n = \underline{n} e \mu_n E \tag{1.18}$$

Since $\sigma = e\mu_n n$, Eq. (1.18) becomes

$$\nabla \cdot I_n = \sigma_n \cdot \nabla \cdot E \tag{1.19}$$

The electric field gradient $\nabla \cdot E$ can be written as

$$\nabla \cdot E = \sigma / \varepsilon\varepsilon_0 \tag{1.20}$$

By using Gauss's law, Eqs. (1.19) and (1.20), we get

$$\frac{\partial \rho_n}{\partial t} = -\frac{\sigma_n}{\varepsilon\varepsilon_0} \rho_n \tag{1.21a}$$

where ρ_n is the extra charge given by $-e\underline{n}$, and

$$\tau_{diel} = \frac{\varepsilon\varepsilon_0}{\sigma_n} \tag{1.21b}$$

the solution of this equation can be written as

$$\rho_n = \rho_{no} e^{-t/\tau_{diel}} \tag{1.22}$$

Thus, the charge neutrality can be eliminated exponentially with a dielectric relaxation time constant, τ_{diel}. The evaluation of this time constant is necessary, because it will play an important role in limiting the speed and response of photodetectors. Let us consider a semiconductor with a resistivity of 10,000 $\Omega\cdot$cm, $\varepsilon = 12$, and $\varepsilon_0 = 8.85 \times 10^{-12}$ F/m. Using Eq. (1.21), $\tau_{diel} \approx 1 \times 10^{-8}$ sec.

For low resistive material the dielectric time constant could be on the order of picoseconds, but for high resistive material it varies from milliseconds to seconds or even longer. This aspect is crucial in examining the response time of the photoconductor.

Here we have discussed the meaning of the term "time constants" as fundamental and useful in understanding photoconductivity and related phenomena. It is important to be aware of the difference between the various time constants even though in some cases it is apparently insignificant.

1.2.18. Noise Current

There are several sources [45] of the random fluctuations in current or voltage that interfere with measurements and are referred to as noise. Noise is an undesirable but unavoidable property of the system and is present in dark current as well as signal current.

Attention should be given to the qualitative and quantitative evaluation of the dark current noise of the photodetector. The noise content provides valuable information about the detection capability and reproducibility of the detection system, and also avoids (or at least estimates) false signals in alarm applications.

In laboratory measurements, it is found that dark noise output is a function of the spectral scan position if a variable filter spectrometer is employed. In this case, it is necessary to carry out a statistical analysis of dark noise, which is done by recording the digitized output for several scans. Sample mean variation and standard deviation as a function of wavelength can be obtained by the usual sttistical methods (standard programs are available for microcomputers). This helps to evaluate reproducibility. A complete account of the methods will not be given here, as an excellent textbook is available [46].

Dark current noise analysis is desirable, but unfortunately it is not customary. An example of the usefulness of dark current analysis is provided by the work of Antreasyan et al. [47]. They carried out a study on photoconductive detectors and obtained useful information that helped them to determine the optimum bias voltage. For all these reasons this aspect should be given proper consideration in a weak signal detection and analysis technique.

1.2.19. Signal-to-Noise Ratio (S/N)

Signal-to-noise ratio is an importnt parameter of a given photosensor; it is expressed in terms of noise in a particular bandwidth. Since the photon counting system is now applied in both laboratories and industry, it is necessary to define this parameter in the following form: Signal-to-noise ratio (S/N) is the ratio of the number of photon counts to the standard deviation for that number in count duration.

Noise plays a crucial role in photoconductivity, particularly in the measurement limits of photosignals. It is useful and significant to know how the magnitude of noise is measured, that is, peak to peak, half-peak to half-peak, average, or root-mean-square (rms) voltage. Unfortunately, this information is rarely given.

1.2.20. Noise Equivalent Power

Noise equivalent power (NEP) is the amount of radiant flux in watts incident on the detector that gives a signal-to-noise ratio equal to 1. It is necessary to specify the chopping frequency at which the measurement is carried out. Noise equivalent power is generally referred to a 1-Hz bandwidth, if not otherwise stated explicitly. The magnitude of a photosignal depends on the wavelength, and therefore the spectral content of the radiation should also be mentioned. Generally, for NEP measurements, radiation from a blackbody source (maintained at 500 K) or monochromatic radiation at the peak of detector response is employed. As mentioned earlier, both photoresponse and noise are a function of temperature, and hence NEP varies with temperature. Therefore, the operating temperature at which NEP is measured should be specified.

1.2.21. Detectivity

Detectivity is the reciprocal of NEP and is measured in reciprocal watts (W^{-1}). Detectivity is the figure of merit of the device and provides the same information as NEP but describes the lower radiation limit to which the photodetector can respond.

It should be borne in mind that the noise under discussion here is average statistical noise. The instantaneous value of the noise frequently exceeds the standard deviation. Since the signal measurement is carried out against the instantaneous random noise, the minimum detectable signal could be about four or five times more than estimated.

1.2.22. Specific Detectivity

There are some types of detectors for which the noise voltage is proportional to the square root of the active area of the photodetector, and hence NEP depends on that area. In these special cases, detectivity needs to be normalized with area and electrical noise band width. For these detectors, specific detectivity is a more appropriate figure of merit than NEP and is defined as

$$D^* = \frac{A^{1/2}(\Delta f)^{1/2}}{NEP} \tag{1.23a}$$

where A is the active area and Δf is the noise band width. Equation (1.23a) can also be expressed as

$$D^* = [RA^{1/2}(\Delta f)^{1/2}]/V_T \tag{1.23b}$$

where V_T is the total noise voltage in a given detector and R corresponds to a peak responsivity.

Responsivity and noise both depend upon the operating temperature, and hence on the specific detectivity also. When the temperature is reduced from 300 K to 77 K, the specific detectivity generally increases by two orders of magnitude. A typical example is shown in Fig. 1.7, where the detectivity of InSb is shown for three different temperatures. Therefore, infrared photodetectors are generally used at low temperatures.

If the detectors are of the extrinsic type, then the operating temperature is much lower—between liquid hydrogen temperature (20 K) and liquid helium temperature (4 K), depending upon the ionization energy of the impurity atoms. Naturally, a sudden change in detectivity is observed near this operating temperature. A typical example is the detectivity of extrinsic detector Ge:Hg, which is shown in Fig. 1.8.

There are some experimental difficulties in calibrating D* for absolute intensities of radiation, and calibration is simplified through the measurement of D*(T) for a blackbody radiation source at a particular temperature, generally 500 K. The spectral peak responsivity R and the specific detectivity D* are related to 500 K blackbody radiation by [18]

$$R_\lambda\,(500\text{ K}) = R_{\lambda 0}(500\text{ K}) \times \text{constant} \tag{1.24}$$

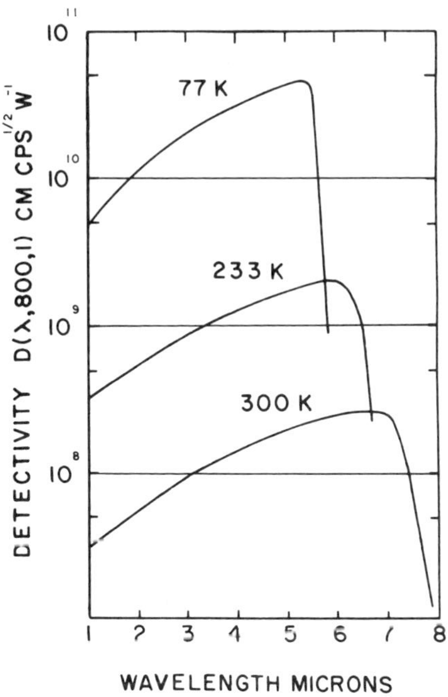

FIGURE 1.7 Variation of detectivity as a function of temperature for a typical indium antimonide photoconductive cell. [After F. D. Morten and R. E. King, Applied Optics 4, 659 (1965)].

and therefore

$$D^*_{\lambda 0}(500\ \mathrm{K}) = D_{\lambda 0} G \tag{1.25}$$

where G is the function defined by Kruse et al. [18].

Noise equivalent power, detectivity, and specific detectivity in a given detector mainly depend on the noise sources and their magnitudes. Theoretical and numerical calculations of these values for several real cases have already been carried out [48], and readers are referred to the original work for details. It is worth pointing out that the magnitudes of these parameters depend upon the type of limiting noise and vary from device to device.

The frequency dependence of detectivity can be expressed as [49]

$$D^*_{rf} = D^* \frac{S/N}{(S/N)_{rf}} \tag{1.26}$$

TABLE 1.1 Parameters, Symbols, and Units Used in Detectors (Visible and Infrared)

Parameter	Symbol	Units	Comments
1. Signal response	S	V or A	The choice of volts or amperes depends on the circuit used for measurements
2. Radiant flux	ϕ	photons/$cm^2 \cdot s$	
3. Responsivity	$R(\lambda)$	A/W or V/W	Lumen is generally not used.
4. Quantum efficiency	η		
5. Dark current	I_d	A	Average value of the dark current. It is desirable to state magnitude of the fluctuation in it.
6. Noise	N	A_{rms}, V_{rms}	Root-mean-square value of current or voltage.
7. Signal-to-noise ratio	S/N		An important parameter of the device. Unfortunately its dependence on modulation frequency, bias voltage, and intensity of radiation is very rarely mentioned.
8. (a) Response time	τ_{res}	sec	Useful for determining speed. It should be specified with the intensity of radiation. Response time can vary by orders of magnitude by varying intensity.
(b) Lifetime	τ	sec	
(c) Dielectric time	τ_{diel}	sec	
(d) Collision time	τ_{coll}	sec	
(e) Transient time	τ_{tran}	sec	
9. Cut of wavelength	λ,10%	Å, μm, nm	Not a significant parameter, but it gives a rough idea of the form of the response curve.
10. Noise equivalent power	NEP	W (watts)	
11. Detectivity	D	W^{-1}	
12. Specific detectivity	D*	$cm \cdot Hz^{1/2}/W$	Figure of merit of a device, dependent on modulation frequency. It may vary with applied voltage.

where D^*_{rf} is the detectivity at reference frequency, which could be a frequency for which the detectivity is maximum. For low-level radiation measurements, the functional dependence of S/N should be obtained in the laboratory. (Manufacturers should provide this information, because it is crucial for low-level signal detection.)

The detectivity (spectral response) of commonly used infrared detectors is given in Figs. 1.9 and 1.10 for intrinsic and extrinsic detectors, respectively. A slight change in the spectral response and in the magnitude of detectivity is observed from manufacturer to manufacturer because of the variation in the quality of the material, impurity content, the technology of introducing impurities, surface condition, and so on. The curves in Fig. 1.9 show the general trend of the spectral response and approximate magnitude of detectivity.

Up to now, we have examined several aspects of Noise Equivalent Power (NEP) and detectivity, both of them essentially depend upon the magnitude of the signal-to-noise ratio. Recent improvements in the digital measurement techniques of photosignal permit suitable processing of noise statistically and increase the detection capability more than expected from the conventional definition.

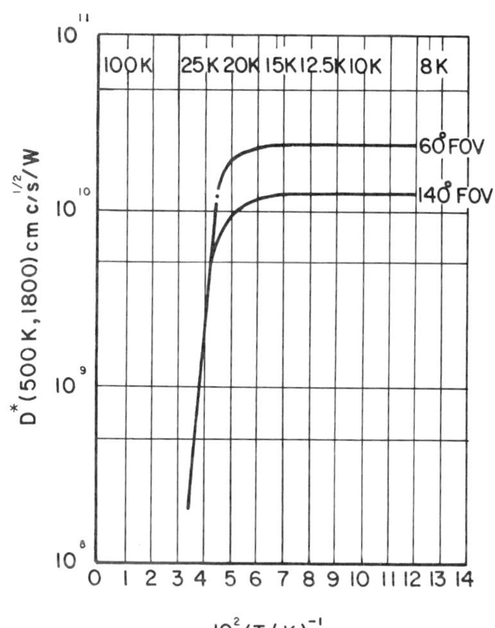

FIGURE 1.8 Variation of detectivity as a function of temperature for extrinsic detectors (Ge:H_g). Experimental results are shown for 60° and 140° fields of view. [After D. E. Bode and H. A. Graham, Infrared Physics 3, 129 (1963).]

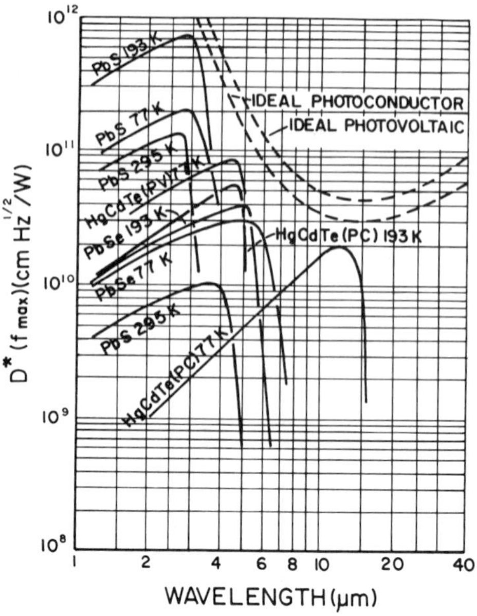

FIGURE 1.9 Detectivity of commercially available intrinsic photodetectors. Note that detectivity is higher at low temperature. [From Photonic Handbook 2, 1985, courtesy of Photonic Spectra magazine.]

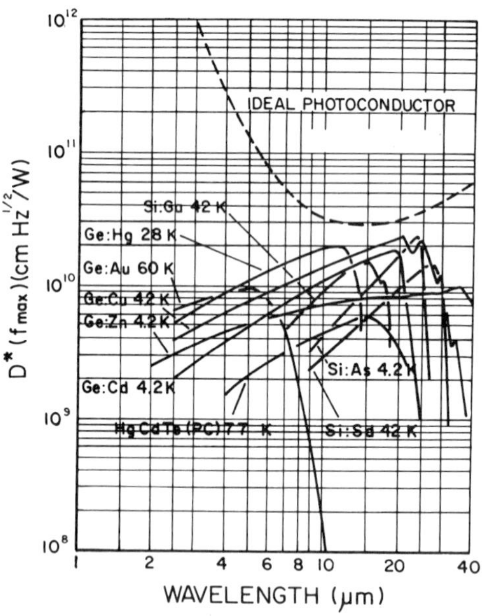

FIGURE 1.10 Detectivity of commercially available extrinsic photodetectors. [From Photonic Handbook 2, 1985, courtesy of Photonic Spectra magazine.]

In the above discussion it is assumed that detector capability is limited by S/N ratio. This is a well accepted fact and is true only when the photosignal is not processed by mathematical/statistical means where the handling of noise is possible. This would help to separate the signal even though $S/N < 1$ and hence, the detection of lower levels of radiation is possible.

To illustrate this point, experimental investigation was carried out on a commercially available CdS Photodetector (Clairox number CL-602) by Serfaty and myself [50]. Steady state photosignals for different intensities and dark current were recorded in transient mode in a digitized form. For maximum value of the photosignal and dark current, a large number of data points were collected and a typical curve is shown in Figure 1.11a. It is clear that in each curve the noise voltage is much higher than the signal and therefore, the presence of the photosignal apparently is not reflected. However, a slight but noticeable shift in the position of the curve is observed when the radiation is incident. The magnitude of the photosignal can be estimated simply by taking the statistical average and subtracting dark current mean value from the mean value of the photovoltage. In this way, the photovoltages for different intensities have been estimated and plotted (Figure 1.11b). As observed frequently, the photocurrent increases linearly with the intensity of the radiation, but in this case, the range of the detected power is much less than the NEP value. Because of its linear nature, the photodetector is obviously useful for low level radiation measurements. The present experimental results demand the revision of the term "NEP" and "Detectivity" on the basis of noise characteristics and its processing rather than using only its r.m.s. value.

1.2.23. Color Temperature Response

The term color temperature response is not frequently used in the literature. The balance of colors is referred to as term "color temperature" [51], because the spectral distribution of the emitted radiation changes with the temperature of an incandescent light source. If, during a measurement, the temperature of the light source is expected to increase (spectral distribution is expected to change), then one needs to search for a detector that has a flat or nearly flat response. If that is not possible, a separate calibration is needed.

This terminology has particular significance when white light needs to be measured quantiatively instead of the monochromatic radiation used in the laboratory. A typical example is the detector used in photographic cameras. For the new programmable cameras, the light measurement system must be very precise. Obviously, a flat response detector, which is adequate for color temperature measurement, is the proper solution.

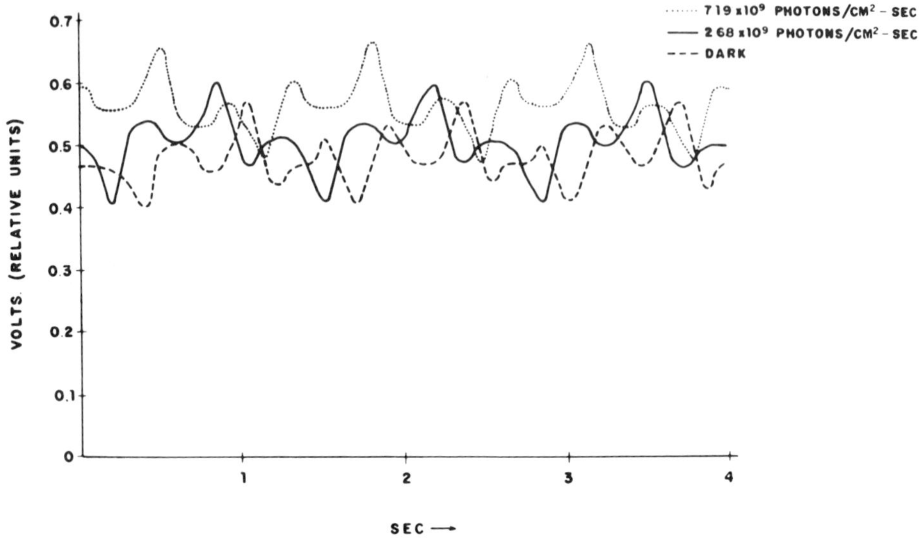

(a)

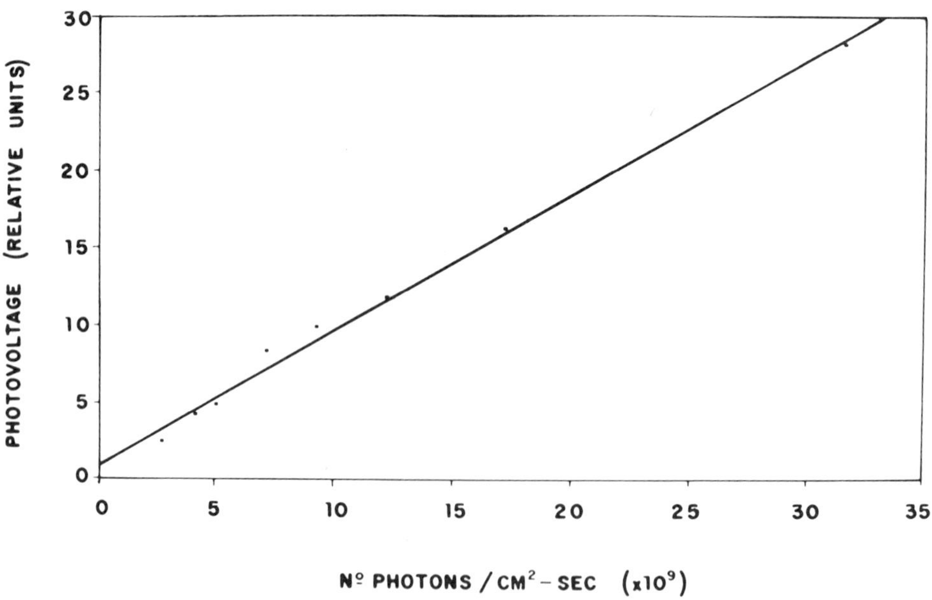

(b)

FIGURE 1.11 (a) Transient form of the steady state photosignal for low intensity of radiation s/N<1. (b) Calculated value of the photovoltage as a function intensity of radiation. (From Ref. 50.)

TABLE 1.2 Color Temperatures of Some Light Sources

Light source	Color temp. (K)
Candle	1900
Tungsten household lamp	2000
Floodlamp	3000
Flashbulb	4000
Daylight (normal sunny day)	5000
Strobe lamp	6000

In order to give a fair idea of color temperature, a list of commonly used light sources and their color temperatures is given in Table 1.2.

To reach the ultimate limit of detectivity is not an easy task. It requires examining the impedance of the detector and special attention should be given to the preamplifier. Some previous work could be a guideline for this purpose. In fact, the level of the signal, properties of the noise, impedance of the detecting circuit, and preamplifier consideration should be carried out in a global form, to achieve the minimum level of detectivity. The proper handling of such information is certainly more an art than a science.

REFERENCES

1. W. Smith, Nature 7, 303 (1873).
2. R. H. Bube, Photoconductivity of Solids. Wiley, New York, 1960.
3. T. S. Moss, Photoconductivity in the Elements. Academic, New York, 1962.
4. H. Levinstein, Appl. Opt. 4, 639, 1965.
5. R. A. Stradling, in New Developments in Semiconductors, P. R. Wallace, R. Harris, and M. J. Zuckermann, Eds., p. 372. Noordhoff International, Leyden.
6. B. R. Pampling, N. V. Joshi, and C. Schwab, Eds. Proc. 6th Int. Conf. on Ternary and Multinary Compounds in Progr. Crystal Growth and Characterization 10, 1984.
7. N. V. Joshi, J. Martin, and P. Quintero, Appl. Phys. Lett. 39, 79 (1981).
8. N. V. Joshi, L. G. Roa, and A. B. Vincent, Nuevo Cimento 2D, 1880 (1983).
9. J. M. Martin, N. V. Joshi, and A. B. Vincent, SPIE Proc. 395, 267 (1983).

10. C. A. Hogarth, Materials Used in Semiconductor Devices. Interscience, New York, 1965.
11. R. H. Bube, in Photoconductivity in Physics and Chemistry of II-VI Compounds. M. Aven and J. S. Prener, Eds., North-Holland, Amsterdam, 1967.
12. T. S. Moss, in Photoconductivity in III–V Compounds in Semiconductors and Semimetals, Vol. 2. R. K. Wildarson and A. C. Beer. Eds. Academic, New York, 417, 1966, p. 205.
13. R. C. Smith, Device applications of the ternary semiconducting compounds. J. Phys. Colloq. 36, C3–89 (1975).
14. S. M. Ryvkin, Photoelectric Effects in Semiconductors. Consultants Bureau, New York, 1964.
15. A. Ambroziak, Semiconductor Photoelectric Devices. Gordon & Breach, London, 1968.
16. A. Rose, Concepts in Photoconductivity and Allied Problems. Interscience, New York, 1963.
17. P. Gorlich, Problems in Photoconductivity. Adv. Electron. Electron Phys. 14, 37 (1963).
18. P. W. Kruse, L. D. McGlauchlin, and R. B. McQuistan, Elements of Infrared Technology. Wiley, New York, 1962.
19. S. Larach (Ed.), Photoelectric Material and Devices. Van Nostrand, Princeton, N.J., 1965.
20. J. B. Dance, Photoelectronic Devices. Butterworths, London, 1969.
21. E. H. Putley, Far Infrared Photoconductivity. Phys. Stat. Solidi 6, 571 (1964).
22. V. Dulin, Electron Devices, MIR, Moscow, 1980.
23. W. Ruppel, Photoconductor-Metal Contacts, in II-VI Semiconductors and Semimetals, D. G. Thomas, Ed., W.A. Benjamin, New York, 1967, p. 1260.
24. T. S. Moss, Rep. Progr. Phys. 28, 15 (1965).
25. R. G. Breckenbridge (Ed.), Proc. I International Conf. in Photoconductivity, Wiley, New York, 1956.
26. H. Levinstein (Ed.), Proc. II Int. Conf. Photoconductivity. Phys. Chem. Solids 22 (1962).
27. E. M. Pell (Ed.), Proc. III Int. Conf. Photoconductivity. J. Phys. Chem. Solids (Suppl.) 32 (1971); also pub. by Pergamon, Oxford, 1971.
28. J. Tauc, Photo and Thermoelectric Effects is Semiconductors. Pergamon, Oxford, 1962.
29. V. S. Vavilov, Effects of Radiation on Semiconductors. Consultants Bureau, New York, 1962.
30. W. T. Tsang (Ed.), Photodetectors in Semiconductors and Semimetals, Vol. 22. Academic, New York, 1985.
31. Y. Marfaing, Photoconductivity and Photoelectric Effects. In: Handbook on Semiconductors, Vol. 2. T. S. Moss, North-Holland, series Ed., Amsterdam, 1980.

32. R. M. Schaffert, Electrophotography. Focal Press, London, 1965.
33. W. F. Berg and K. Hauffe (Eds.), Current Problems in Electrophotography. de Gruyter, Berlin, 1972.
34. J. Mort and C. M. Pai, Photoconductivity and Related Phenomena. Elsevier, New York, 1976.
35. J. Schmit, in Semiconductors and Semimetals, Vol. 5, R. K. Willardson and A. C. Beer, Eds., Academic, New York, 1970, p. 175.
36. P. Kruse, in Semiconductors and Semimetals, Vol. 5, R. K. Willardson and A. C. Beer, Eds., Academic, New York, 1970, p. 15.
37. International Optical Computing Conferences 1 and 2. Proc. SPIE 231 and 232 (1980).
38. Lynn D. Hutcheson, Integrated Optical Circuits and Components. Marcel Dekker, New York, 1987.
39. G. A. Mourou and D. M. Bloom, Eds. Picosecond Electrons and Optoelectronics. Electrophysics Series, 21, Springer Verlag (1985).
40. V. D. Mattera, Jr., F. Capasso, J. Allam, A. L. Hutchinson, J. Dick, J. M. Brown, and A. Westpal, J. Appl. Phys. 60, 2609 (1986).
41. H. Temkin, J. C. Bean, T. P. Pearsall, N. A. Olsson, and D. V. Lang, Appl. Phys. Lett. 49,155 (1986).
42. J. L. Shay and J. H. Warnick, Ternary Chalcopyrite Semiconductors : Growth, Electronic Properties and Applications. Pergamon, Ne York, 1975.
43. N. V. Joshi, L. Martinez, and R. Echeverria, J. Phys. Chem. Solids 42,281 (1981).
44. J. R. Waldram, The Theory of Thermodynamics. Cambridge Univ. Press, Cambridge, U.K., 1985.
45. A. Van der Zeil, Solid State Physical Electronics. Prentice-Hall, Englewood Cliffs, N.J., 1971.
46. C. Wyatt, Radiometric Calibration: Theory and Methods. Academic, New York, 1978.
47. A. Antreasyan, C. Y. Chen, and P. A. Garbinsk, Electron. Lett. 21, 1136 (1985).
48. S. Nudelamn, Appl. Opt. 1,627 (1962).
49. S. M. Sze, Physics of Semiconductor Devices. Wiley-Interscience, New York, 1983.
50. N. V. Joshi and A. Serafty, International Jour. of Infr. and Millimeter Waves, 10, 1077 and 1090, 1989, (in press).
51. Editors of life time books, Color. Time Life International, The Hague, The Netherlands, 1970.

2

Techniques for Photoconductivity Measurements

2.1. INTRODUCTION

To describe adequately the general trend of experimental methods of photoconductivity measurements and their details is not a simple issue, as different laboratories have their specific systems which vary considerably. Here we will not refer to the different methods used in preparing photoconductors or to the various methods for making ohmic or rectifying contacts. To make an "adequate" or "proper" contact to photoconductors is not a simple problem and there is not an easy solution. This is because the contact material and contact techniques depend on the semiconductor itself. Moreover, both parameters (material and the techniques for making electrical contacts) depend on the type of majority carriers and very often on the concentration of charge carriers also.

There exists a large amount of general literature on contacts and their effects [1–5]. After reviewing the extensive literature on metal-semiconductor contacts and being struck by its abundance, I find it very difficult to formulate any conclusion or guideline for making adequate contacts to photoconductors. From the point of view of an experimentalist, it is necessary to make several trials and/or to search for an adequate reference for a specific material if such information is available for making ohmic contacts. (For example, the technical details for InP, n or p type, are given in Ref. 6.) Moreover, there exists a literature guide [7] for obtaining information quickly about a particular type of contact for the semiconductor in question. Care should be taken to avoid structural changes that might take place during the process of making contacts to semiconductors, such as the effects observed when gold-based contacts have been made to gallium phosphide [8]. It is worth mentioning that excessive noise in the measurements and non-reproducibility in the finer details of the photoconductivity spectrum are generally the result of bad or improper contacts, and therefore it is necessary to evaluate the electrical contacts before making photoconductivity measurements.

A vast amount of literature [7] suggests that the preparation of semiconductor surfaces for making ohmic contacts is also important. It is much more effective to make in situ ohmic contact to a freshly cleaved surface in ultrahigh vacuum than to a contaminated surface, as adsorbed gases also have an effect on the contacts. However, as it may not be practicable to have a freshly cleaved surface, chemical etching is an alternative solution. The etching reagent depends on the semiconductor, the type of majority carriers, and the crystallographic planes. In this case also, as in the case of contacts, a proper etching reagent should be selected from the literature guide [7]. Surface preparation is important not only for achieving good metal–semiconductor contacts but also for avoiding or reducing the effects of surface states.

Over the past decade much attention has been given, both from the theoretical and experimental [9–13] points of view, to the contribution of semiconductor surfaces to the overall valence band structure. There is abundant evidence from photoemission [14,15] and ultraviolet spectroscopy [16] that the surface energy state distribution overlies the bulk valence state distribution, particularly near the top of the valence band, up to the Fermi level. This is the reason the photosignal is not very sensitive to the surface conditions in chalcogen materials (Se or Te) where surface states are encountered much less frequently. The influence of surface states on the absorption spectrum of CdS has also been reported [17]. Evidently, similar results are expected in photoconductivity, and recent experimental results confirm thsi [18]. Surface preparation has significant consequences with respect to the finer details of the photoconductivity spectrum. Obviously, it has a considerable effect on the optical and electrooptical properties of materials.

The choice of contact material and method of making the contact depend upon the specific requirements. Some types of contacts are desirable for one purpose but seriously harmful for other purposes. This is also true of the geometry of the contacts. For high-speed photoconducting sensors, Schottky barrier type contacts are desirable to increase the collection efficiency of the electrons and to improve the speed of the devices [19]. On the other hand, the same types of contacts pose the problem of analysis of the transient curve [20,21]. In high-speed photodetectors, the distance between two contacts should be as small as possible to reduce transit time, whereas for analysis of the transient curves, the distance between two electrodes should be much greater than the contact regions. Criteria for eliminating the effects of neutral contacts and their pertinent details have already been discussed extensively [21]. Sometimes good, reliable, ohmic or rectifying contacts are not enough, and additional experimental precautions such as contact configuration or a specific separation between contact region and illuminated region are required for photoconductivity (particularly time-dependent) measurements and their analysis. Obviously, contact geometry should be selected appropriately. Contacts on reasonably small materials with a close separation between them (on the order of a few micrometers) are rather uncommon, and processing technology for making them is available in only a few materials (such as GaAs, Si, and Ge). If the contacts are injecting, then the situation is more complex, because this type of contact modifies several properties. First, electron concentration is changed and diffusion currents are set up. In selecting type and geometry of contacts, generally a compromise is made between ideal and practical choices.

2.2. CONVENTIONAL EXPERIMENTAL SETUP AND ITS ANALYSIS

To start, let us first consider the simplest and most conventional transversal arrangement (Fig. 2.1), which can be used for unipolar as well as ambipolar charge carriers. We assume that contacts are ohmic. Contacts should not limit or obstruct the charge carriers and must supply the necessary number of electrons to maintain the neutrality of the sample. Let R_d and R_{Ill} be the resistance of the sample in the dark and under uniform illumination conditions, respectively. The change in resistivity is linear with the intensity of illumination. Photocurrent is generally measured across a load resistance R_L introduced in the biasing circuit. I say "generally" because often, particularly when the resistance of the sample is high (10^5 Ω or more), photocurrent measurements are carried out with a high impedance electrometer as will be discussed later.

Photovoltage across the resistance R_L can be calculated very easily by evaluating the difference between the voltages with and without radiation. When no radiation is incident on the sample voltage is given by

$$V_d = \frac{R_L V}{R_L + R_d} \tag{2.1}$$

Similarly, the voltage for an illiminated sample is given by

$$V_{Ill} = \frac{R_L V}{R_L + R_{Ill}} \tag{2.2}$$

where V is the applied voltage. The difference between them is the signal response, which is given by

$$S \text{ (volts)} = R_L V \left[\frac{1}{R_L + R_{Ill}} - \frac{1}{R_d + R_L} \right] \tag{2.3}$$

$$= R_L V \left[\frac{\Delta R}{(R_L + R_{Ill})(R_d + R_L)} \right] \tag{2.4}$$

where $\Delta R = R_d - R_{Ill}$.

In the laboratory, one generally handles the values of the resistance in the dark, load resistance, matching impedance, and variation of conductance of the sample due to radiation, and therefore it is necessary to convert Eq. (2.4) to make all these terms visible.

Conductance G (in ohms^{-1}) is defined as the reciprocal of the resistance, and therefore ΔG, the difference in the conductance caused by the radiation, is given by [22]

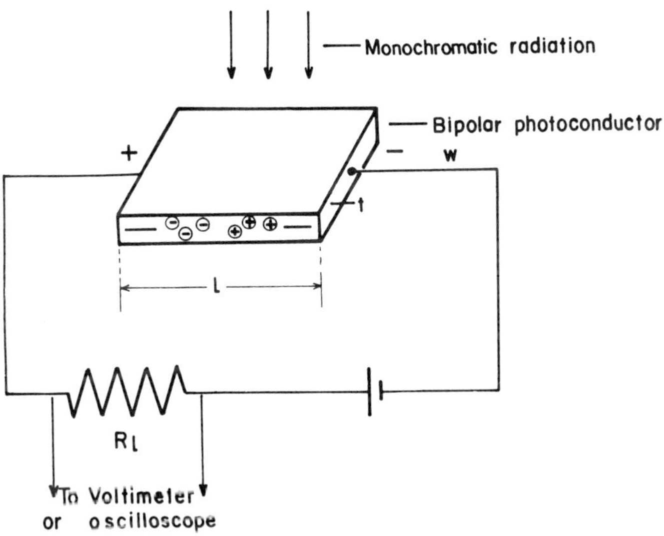

FIGURE 2.1 A conventional experimental setup for photoconductance measurements.

$$\Delta G = \frac{1}{R_{Ill}} - \frac{1}{R_d} \tag{2.5}$$

Therefore,

$$\Delta R = \Delta G\ R_d R_{Ill} \tag{2.6}$$

and hence Eq. (2.4) becomes

$$S \text{ (in volts)} = \frac{V R_L R_d R_{Ill}\ \Delta G}{(R_L + R_d)(R_L + R_{Ill})} \tag{2.7}$$

Here we have assumed that the sample is illuminated uniformly, so that the variation in the conductance is uniform in the plane on which the radiation is incident. However, absorption cannot be uniform in the direction perpendicular to the plane, as the number of photons absorbed varies according to Lambert's law of absorption:

$$\underline{I}(x) = \underline{I}_0 \exp(-\alpha x) \tag{2.8}$$

where $\underline{I}_0$ is the number of incident photons, α is the absorption coefficient, and x is the distance measured from the surface. Thus, the amount of radiation absorbed at each thin layer is different, and therefore the conductance is different. The total variation can be obtained by integrating over entire thickness of the sample

$$\Delta G = \frac{W}{L} \int_0^{\text{thickness}} \Delta\sigma(x)\ dx \tag{2.9}$$

which can be estimated, as all the necessary parameters are known.

Let us consider two limiting cases (these are not hypothetical situations—very often an experimentalist encounters them in the laboratory).

1. Variation in the resistance with and without radiation is very small compared to the dark resistance R_d, that is, $R_d \simeq R_{Ill}$. Equation (2.7) becomes

$$S(V) = \frac{V R_L R_d^{\,2}\ \Delta G}{(R_L + R_{Ill})^2} \tag{2.10}$$

All the parameters expressed in Eq. (2.10) are characteristic of a given photoconductor except load resistance R_L. To a detector user, it is the only parameter that can be controlled to obtain the maximum photoresponse. The value of R_L that gives maximum response can be obtained by using $dS/dR_L = 0$ and is found to be $R_L = R_d$. This is the reason experimentalists use the load resistance equal to the dark resistance of the sample when the photosignal is weak.

2. In the case of high-resistivity material the dark current is low, which makes it very easy to measure photocurrent (or voltage) without any need to optimize the load resistance. On the contrary, if $R_L = R_d$, then the response time increases as a consequence of the circuit time constant ($RC = \tau$). As $R_L \ll R_d$, Eq. (2.7) can be written as

$$S(V) = VR_L \; \Delta G \tag{2.11}$$

The above analysis is valid as long as contacts are neither injecting nor blocking. For analysis of these types of contacts, excellent books and review articles are available [e.g., 7,21]. Here it is also assumed that photogenerated charge carriers are uniformly distributed throughout the sample.

This analysis is frequently used, even though it is an approximate one. Diffusion of charge carriers and details of the recombination processes also have to be taken into consideration.

2.3. ANALYSIS OF CONTACT CONFIGURATION

Photoconductivity is one of a few topics where experimental setup, contact configuration, preparation of the sample, and sample thickness play crucial roles in the output and in the analysis of the results. The overall structure of the spectral response curve may be considered to trace faithfully the physical mechanisms (optical absorption and transport) of the detector, but the magnitude of the photoresponse and the transient properties such as speed and lifetime of the charge carriers may be strongly influenced by experimental details that may appear to be unimportant.

Figure 2.2 shows commonly used contact configurations and the manner in which the sample is illuminated. In Figs. 2.2a and 2.2b the contacts are on the lateral side of the sample (or extreme ends of the illuminated face); but in the latter, only part of the sample is illuminated. In Fig. 2.2c the contacts are on the front and back sides of the sample. The radiation passes through the transparent contact and is nonuniformly absorbed in the material. The amount of absorbed radiation varies exponentially with the depth. Obviously, in this case, photogeneration, recombination, and there-

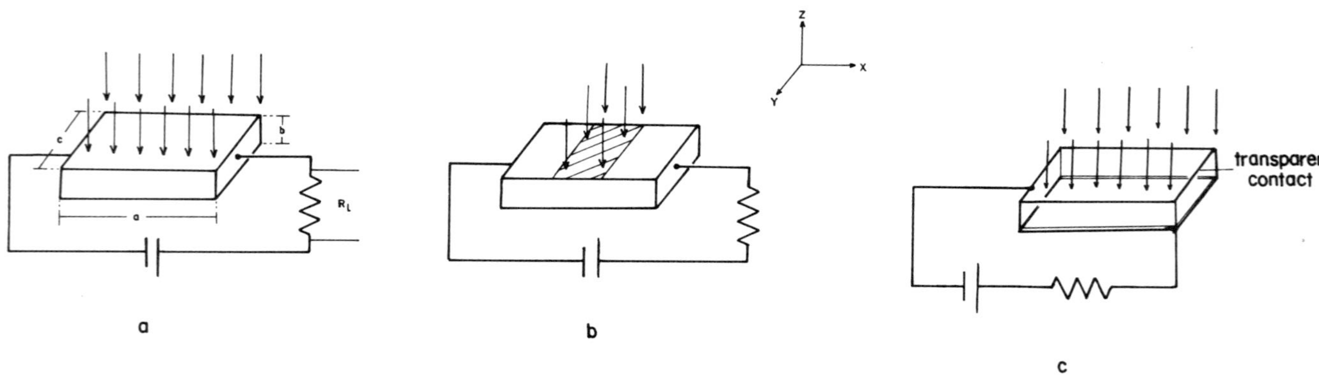

FIGURE 2.2 Different types of contact configurations that dramatically alter the assumptions involved in the analysis of the results.

fore, photoconduction depend on the thickness of the sample. Front surface recombination plays an important role in the first two cases (Figs. 2.2a and b), whereas the back surface recombination is also important for the last case.

Unless otherwise mentioned, it is assumed that the beam profile is uniform and the sample is homogeneous.

The calculations of the magnitude of the photoresponse depend on several approximations, which vary with the configuration and experimental setup. It is therefore necessary to consider limitations, approximations, and mathematical formalism for each case separately.

Let us consider a frequently used configuration (see Fig. 2.2a) in which the sample is illuminated uniformly by radiation having photon energy higher than the band-gap energy. As the absorption coefficient is quite high, the radiation is completely absorbed near the surface, and hence a greater excess of charge carriers are created in this region. Because of the difference in the densities at the illuminated and nonilluminated sides of the sample, the diffusion of the excess charge carriers takes place from the former to the latter. In the presence of an electric field, the diffusion equation can be given by [23]

$$\frac{\partial \underline{P}}{\partial t} = D_P \nabla^2 \underline{P} - \frac{\partial}{\partial x}(\mu_P \underline{P} E) - \frac{\underline{P}}{\tau} \tag{2.12}$$

Equation (2.12) in the three-dimensional case is written as

$$\frac{\partial \underline{P}}{\partial t} = D_P \left(\frac{\partial^2 \underline{P}}{\partial x^2} + \frac{\partial^2 \underline{P}}{\partial y^2} + \frac{\partial^2 \underline{P}}{\partial z^2} \right) - \mu_P E_x \frac{\partial \underline{P}}{\partial x} - \frac{\underline{P}}{\tau} \tag{2.13}$$

As the radiation is uniform in the xy plane, the terms $\partial^2 \underline{P}/\partial x^2$, $\partial^2 \underline{P}/\partial y^2$ can be neglected. Similarly, the driving electrical field is also kept small, and therefore $\mu p E_x\ \partial \underline{P}/\partial x$ is not considered. Equation (2.13) then becomes

$$\frac{\partial \underline{P}}{\partial t} = D_P \frac{\partial^2 \underline{P}}{\partial z^2} - \frac{\underline{P}}{\tau} \tag{2.14}$$

Here D_P is the diffusion constant for holes and is equal to $\mu_p kT/e$. In obtaining this expression it is assumed that the particles are free and there is not interaction between them. This is true only when the density of the charge carrier is low.

Equation (2.14) is frequently used for estimating the time and space dependence of photogenerated charge carriers. However, in the last decade, the use of a high-power source, such as a laser beam, is very common, and in this case spatial concentration of photogenerated charge carriers is created. This makes the diffusion process nonlinear; that is, the differential equation (2.14) should contain higher-order terms. In this situation the term "diffusion coefficient" is also ill-defined. In short, Eq. (2.14) loses its applicability when the incident power is very high. Since this is a partial differential equation, its solution depends on the boundary conditions, which in turn vary with the ratio of the thickness b over the diffusion length L_h, which is equal to $[D_p\tau]^{1/2}$. The solution of Eq. (2.14), and hence variation in the conductance, has already been worked out [23] for two relative values of thicknesses compared to diffusion length and is given by

$$(\Delta G)_0 = \frac{ce(\mu_n + \mu_P)}{a} g_0\left[\frac{1}{\tau} + \frac{V_S}{L_h}\right] \tag{2.15}$$

for b/L > 1, and

$$(\Delta G)_0 = \frac{\beta\psi)\mu_n + \mu_P)g_0\, e\; \frac{D_P}{L_h V_S} + \frac{b}{2L_h}}{aV_S\{[1 + (D_P/L_h V_S)^2](b/L_h) + 2D_P/L_h V_S\}} \tag{2.16}$$

for b/L < 1. Here V_S is the surface recombination rate.

Depending upon the relative values of D, V, and L_h, an approximate solution can be obtained for a particular case. The equations show the importance of the ratio b/L for estimating the magnitude of photoconductance for the configuration shown in Fig. 2.2a.

This treatment is an extension of the one described in Section 2.2 and takes into account the diffusion of the photoexcited carriers and surface recombination. In this configuration, the contacts are illuminated, and their effects merit separate consideration.

The role of the contacts (even unilluminated) to the photoconductor, independent of their nature (blocking, injecting, or neutral), is also a complex issue. In photoconductivity measurements, the response depends not only on the number of photogenerated charge carriers but also on their transport to the electrodes. While the contacts are made between metals and semiconductors, depending upon the work function, there is a flow of charge carriers from the higher to the lower Fermi level to equalize the Fermi level at the contact. In this process, electrons can transfer from the semiconductor to the metal, resulting in a depletion region in the semiconductor. This gives rise to perturbation of the band at the metal—semiconductor interface. The flow of charge carriers is influenced by the length of the depletion region and the amount of perturbation at the interface.

When the regions near the contacts are illuminated, quasi-Fermi levels for electrons and holes move relative to the Fermi level of the metal. If the illumination is intense, the movement of the Fermi level may be large enough to change the nature of the contact and hence change its overall properties. It is, therefore, possible to find a substantial change in the photoconducting properties if the contact region is illuminated.

Now let us examine the experimental situation shown in Fig. 2.2b, where only the central part is illuminated. This configuration has additional complications. First, the diffusion is not only in the z direction but also in the x direction, and, second, the band structure is distorted at the boundary of the illuminated area as quasi-Fermi levels for electrons and holes in the illuminated portion shift with respect to the unilluminated sample.

The effects of diffusion in the x direction can be taken into account rather easily. The excess charge carriers are expected to extend over a diffusion length $(D_p \tau_p)^{1/2}$ on both sides of the illuminated portion. If the length of the illuminated region is much larger than the diffusion length, then end effects are small and can be neglected [23]. Similarly, there is a limit for the magnitude of the applied voltage within which the diffusion effect is very small. This calculation has been provided elsewhere [23]. If both requirements are fulfilled, then the discussion and results of Fig. 2.2a could be roughly extended to the situation depicted in Fig. 2.2b.

In the configuration of Fig. 2.2b, what is more complicated and difficult to handle are the effects caused by the internal field created when Fermi levels of illuminated and unilluminated portions are made to coincide, creating a local potential barrier. The calculation details of this internal field effect are given by Tauc [24]. Such an effect is predominant in ferroelectric materials [25].

The other consequence of this configuration in high-resistivity materials is the creation of space charge in the illuminated region. This is understandable because in these materials the concentration of deep trapping centers is quite high, and a reasonable fraction of photoexcited charge carriers are trapped. Under these circumstances, the charge is uniformly distributed over the illuminated region, which screens the applied voltage. The magnitude of the altered voltage depends upon the size and shape of the illuminated region with respect to the electric field. For the elliptical form of the spot, some calculations have been carried out by Uyukin [26]. In the presence of screening, the lifetime of the charge carriers and hence the photoresponse is affected.

It is worth mentioning that such effects are of particular importance when the intensity of the radiation is high and the conductances of the illuminated and unilluminated portions differ considerably. Special attention should be given to this aspect.

Now consider the third and very frequently observed situation where light penetrates through a transparent contact. The charge carriers are created in the entire material and collect on the back of the device. In Chapter 6 we will see some devices based on this configuration. In these cases, almost all the carriers are generated in a thin layer of thickness $1/\alpha$ and diffuse toward the back of the sample. When the thickness is much greater than the carrier diffusion length, then the carrier distribution inside the sample is exponential and at the back the concentration is very small, near zero. When the sample thickness is comparable to the diffusion length, the situation is different. Concentration at any point depends upon the generation rate, the recombination rate at the back side, and the diffusion of the carriers from the front side. A typical distribution of charge carriers as a function of thickness is shown in Fig. 2.3.

In this case, the continuity equation is similar to Eq. (2.14) but the generation rate of charge carriers varies with the depth z, and this should be taken into account. Then for an n-type photoconductor, Eq. (2.14) becomes

$$D_p \frac{\partial^2 \underline{p}}{\partial z^2} - \frac{\underline{p}}{\tau_1} = g(z) \tag{2.17}$$

where τ_1 corresponds to the time constant related to the recombination process. Solutions of this type of equation depend on the boundary conditions. For Z = b (i.e., on the front surface), the current is due to surface recombination and is given by $V_{S\underline{p}}$. The boundary conditions on the back surface, however, vary with the thickness of the sample. Hence, three distinct situations are possible according to the relative value of b with respect to the diffusion length L_h. Thus Eq. (2.17) needs to be examined for the three cases $b \gg L$, $b = L$, and $b \ll L$.

The solution of Eq. (2.17) gives information on $\underline{p}$ as a function of z for a given set of boundary values. The number of photogenerated charge carriers can be estimated by integrating the distribution function p(z) derived from Eq. (2.17) over the thickness b of the sample and is given by

$$\underline{p}(z) = \int_0^b \underline{p}(z)\, dz \tag{2.18}$$

Boundary values (selected or approximated) are crucial in obtaining the solution of partial differential equations.

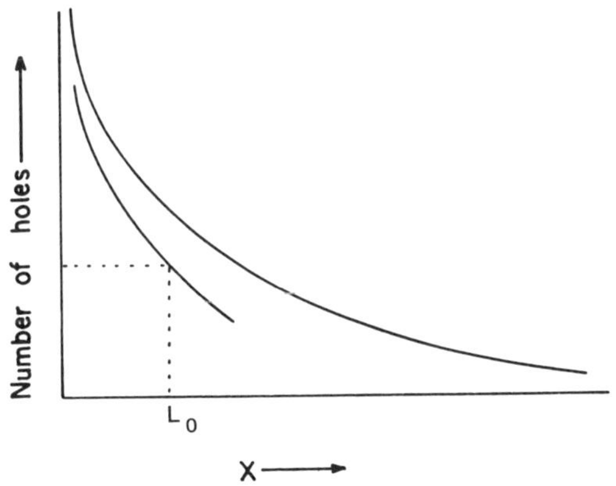

FIGURE 2.3 Typical curves showing the number of holes as a function of depth of penetration. The form varies according to the boundary conditions and the velocity of recombination at the surfaces.

Some typical calculations have been provided by Moss [27] and by Ambroziak [28] for commonly encountered situations. For a specific device, a detailed calculation is necessary to estimate the sensitivity of other parameters. For the majority of devices, it is not possible to obtain an exact solution of differential equations for a specific set of boundary conditions. In such circumstances, adequate approximations are used or numerical methods can be employed.

The purpose of the present discussion is not to suggest any particular configuration or any approach to handle the complications but to bring to notice the role of several parameters and their effectiveness. The experimentalist should be aware of the approximations and their side effects within a particular configuration. It is really unfortunate that only a small percentage of published papers have mentioned the details of contact configuration, surface conditions, and experimental setup with which the results were obtained.

2.4. METHODS OF MEASUREMENT

When the response of a given sensor is reasonably high, then photoconductance measurements are really simple provided that the signal to be detected is not too weak. However, for several reasons one needs to measure photoresponse and electrooptical properties of semiconductors that show a poor photoresponse, and in that case it becomes difficult to obtain a reliable photosignal. Such information is valuable not only for understanding the band structure parameters, but also for knowing the presence of optically active centers, whether donors or acceptors.

The photoconducting properties of a given semiconductor depend on several electrical and optical parameters, and many of them vary considerably. For example, electrical conductivity could vary by 10 orders of magnitude, and the optical absorption coefficient by four orders of magnitude. Obviously, an experimentalist has to be prepared to face a variety of experimental situations. The semiconductor can have high or low resistance, the magnitude of the variation in the resistance can be substantial, low, or negligible. The response time can vary from a few picoseconds to a few minutes (very rarely up to hours). Even though theory and analysis for the observed photoconductivity spectrum are basically the same in all cases, the experimental technique must be different for each case.

Therefore, a few experimental methods for measuring steady state photoconductance are discussed below. Only some of them are commonly used in various laboratories. They are classified according to the characteristics of the photoconductors that make them suitable for use with a specific method. I also want to bring out why a particular method is more suitable than others in certain conditions.

Case 1. High-Resistivity and Good Photoresponse

These types of materials obviously have low dark current I_d, and their photocurrent is much higher than I_d. The measurements on such systems are always easy to carry out, and the experimental details are given in Fig. 2.1. This is the simplest and easiest situation for recording photosignals (signal may be in current or in voltage).

In order to study photoconductivity response curves, one needs to examine the photoresponse as a function of the wavelength, that is, the measurements should be carried out over the entire spectral range. Obviously, the radiation source should be continuous in wavelength. The selection of a proper source depends upon the spectral region and will be discussed separately.

If the relaxation curves are to be analyzed, then the voltmeter should be replaced with an appropriate oscilloscope. The basic information and several forms are discussed by Ryvkin [21]. The most important aspect to remember is that the RC time constant of the measuring circuit should not be a limiting factor for the relaxation rise or decay curve. Such an error can occur because the RC constant can vary from 10^{-6} to 10^{-1} sec, and this value sometimes coincides with the actual value of the response time. In wide-band-gap semiconductors, particularly group II–VI compounds, the response is slow and the above-mentioned situation is very common.

Case 2. Low-Resistivity and Reasonably Good Photoresponse

This is a very common situation in narrow-band-gap materials ($E_g \leq 1$ eV). Dark current is of the order of microamperes or even higher. In this case, of course, the above method cannot be used and the modulation technique should be employed. The basic principle is shown in Fig. 2.4. The incident beam is chopped with a certain frequency by a mechanical or electrooptical (Kerr cell) chopper. This makes the photocurrent periodic with frequency equal to that of the chopper. The output now is an ac photosignal on dc dark current. The ac signal, even though very weak, can be separated by an ac or phase-sensitive amplifier. This is a very useful and widely used technique, particularly in infrared detectors.

This method has the additional advantage that it helps to reduce the noise, which in photoconductors is an important aspect for detecting weak signals. Separate consideration will be given to this in Chapter 5.

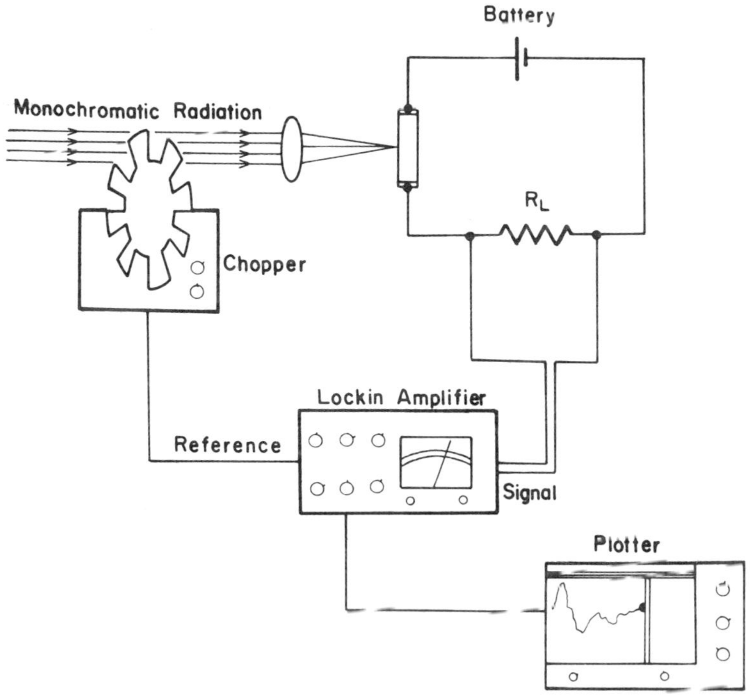

FIGURE 2.4 Phase-sensitive photocurrent measurement system. A photosignal buried in the dark current can be separated easily.

The frequency of modulation of the light beam is determined in such a way as to match the response time of the photoconductor. If the response time is slow, then it takes a long time to attain steady state photocurrent, that is, the frequency should be low. Similarly, if the photoconductor has high noise in a certain frequency range, then that range should be avoided. This technique was used successfully in recording the photoconductivity spectrum of the ternary compound $AgInS_2$ [29].

Case 3. Low Resistance and Slow Response

This is one of the difficult situations where photodetection, particularly of weak signals, is not easy. The value of the dark current is quite high, much higher than that of the photocurrent, and the chopping technique is not useful because of the slow response. In this case, dark current can be balanced by a Wheatstone bridge, and then the potential difference originating from the variance of the resistance of the sample can be amplified by an operational amplifier. In this way the proper impedance at the entrance of the nanovoltmeter or equivalent instrument can also be achieved. Experimental details are given in Fig. 2.5. This technique has been used to detect weak signals in the ternary compound $CuInTe_2$ [30].

If the dark current is comparable to the photocurrent, then this method is not necessary, since zero suppression of normal sensitive equipment helps to separate the photosignal with relative ease.

Case 4. High Resistivity, Slow Response, and Very Weak Photosignal

This is not only a very common situation with organic semiconductors [31,32] and polymers but is also frequently encountered with inorganic semiconductors. Typical examples are the base materials

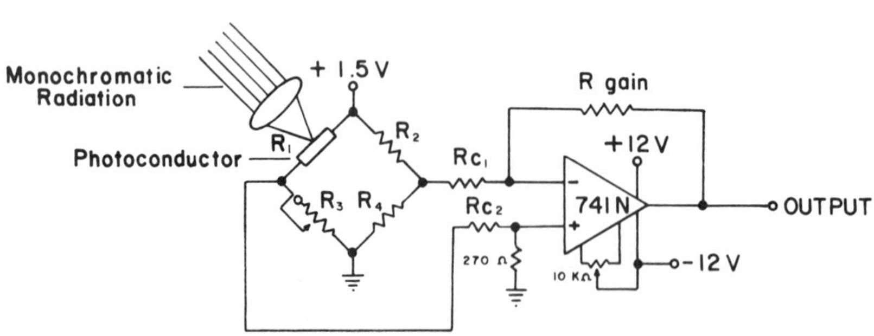

FIGURE 2.5 Wheatstone bridge with a typical amplifying circuit used to detect a weak photosignal that has a slow response. [After Ref. 30.]

used for integrated circuits or for electrooptical devices such as chromium-doped semi-insulating GaAs or iron-doped InP. The resistivity of such samples is high, and the response is slow. This is because the impurity atom creates defect states, which may act as recombination centers. The recombination process reduces the lifetime of the charge carriers in the conduction or valence band, and therefore the intensity of the response is reduced. In addition, the traps retain electrons for a considerable time and hence the response becomes slow (seconds or even a few minutes). The conventional phase-sensitive technique, obviously, cannot be employed. The only choice is to carry out the measurements point by point; for example, for every wavelength, sufficient time should be given for the photocurrent to reach its saturation value. In this manner, the photoconductivity spectrum of chromium-doped GaAs has been recorded by allowing 1 hr for stabilization of the photocurrent. Such a study was necessary to reveal the structural details [33]. This is a tedious task, which microprocessers help to carry out in a routine manner. A recently developed technique will be discussed below.

2.4.1. Step-by-Step Scanning Techniques

It is well known that in several photoconductors the rise time consists of two predominant components as shown in Fig. 2.6. The fast component could be about 10 or perhaps 100 times faster than the slow one, and the contribution of the slow component is significant. In that case, conventional methods, such as the radiation modulation technique or any of the above-mentioned methods, provide information and include only the fast contribution. No attempt is made to take into account the slow contribution, which is very important to understanding the role of the traps and sometimes reveals the de-

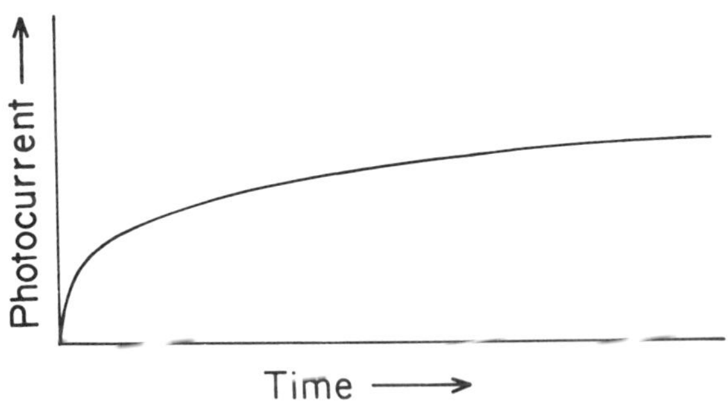

FIGURE 2.6 Time dependence of photocurrent indicating a substantial contribution from the slow component.

tails of their energy distribution. Barreto et al. [34] have improved the conventional methods. Instead of scanning the photoconductivity spectrum continuously, they designed a special computer-controlled experimental setup that enables step-by-step scanning and allows the photoresponse to be measured for a desired amount of time at each wavelength. The experimentl setup reported is shown in Fig. 2.7.

The movement of the monochromator is controlled by a step motor with the help of a microprocessor. It is convenient to use a parallel interface IEEE 488 to communicate with the step motor. Controlling signals were generated through the gates PC_1 and PC_2, amplified through proper circuitry, and fed to the step motor. One signal consists of a train of 50 pulses per second, which set a monochromator at the desired wavelength for the required period of time (a few minutes to a few hours). The other signal controls the backward or forward motion. With a proper program (preferably in machine language for quick and easy control), the position of the monochromator can be controlled with the precision of a fraction of an angstrom with excellent reproducibility. This system is found to be just right for scanning step by step and for radiating the sample for the desired (and necessary) amount of time.

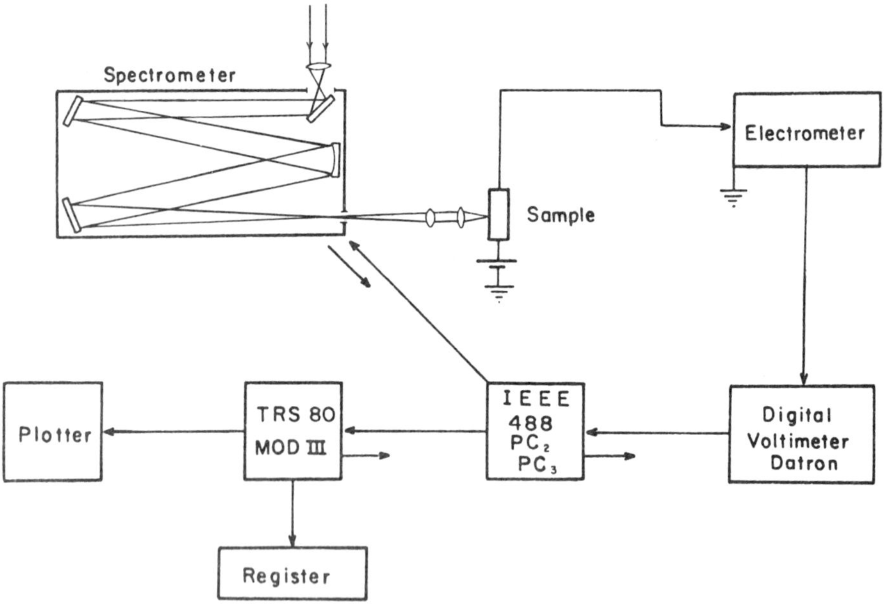

FIGURE 2.7 A computer-controlled system for photodetection [34].

Interface IEEE 488 also controls the measurement of the digitized photosignal (either in the current or the voltage mode). The photoconductivity spectrum recorded with this technique permits enough time to be given to collect the photoresponse from the slow component. In addition to this, the system permits repeated measurement of the same photoresponse so that the mean value can be used, thus simplifying further mathematical and statistical operations such as normalization and taking the first and second derivatives. In short, data handling of photoresponse becomes much simpler and direct. Obviously, the present method shows superiority over conventional methods where the monochromator scans the spectrum continuously.

Recently, by use of this technique, photoconducting transitions of the ternary semiconducting compound $ZnSiP_2$ have been revealed. Similarly, some finer details of the photoconductivity spectrum of a commercially available CdSe:Cu detector were reported [34].

2.5. TIME-DEPENDENT PHOTOCONDUCTIVITY MEASUREMENTS—IMPORTANCE AND POTENTIAL APPLICATIONS

When the sample is illuminated, it takes a certain time for the photoresponse to reach its maximum value. The variation of photoconductance as a function of time over that duration is called transient photoconductivity.

Recording and analyzing [35] transient photoconductivity on oscilloscopes (or on recorders) has been practiced increasingly over the last few years. Conventional techniques provide only qualitative and approximate information about the kinetics of the transient phenomenon. Joshi and Martin]36] recorded with precision the time dependence of photocurrents in digitized form and obtained useful information about the delayed component. Now time-dependence study can be used for investigating

Response time
Discrimination between two different rise time constants
Presence of delayed components
The form of the relaxation curves in relation to trap contributions, photomemory effect, and so on

The methods mentioned in Section 2.4 are appropriate for steady state photoconductance. For transient analysis, some modifications in the measurement system are needed.

The response time can be measured with the conventional phase lag method [21]. If the incident beam is chopped with a sinusoidal modulation frequency ω, then the phase difference ϕ between the exciting signal and the alternating component of the photoconducting signal is given by

$$\phi = \tan^{-1} \omega\tau_{res} \tag{2.19}$$

From Eq. (2.19), it is clear that if ω and ϕ are known, the response time can be calculated.

For phase lag measurements, the experimental setup is very simple. A sinusoidal modulated beam is divided in two by a beam splitter. One part is allowed to fall on the sample, and the other on a fast photodetector (preferably with a response time of the order of picoseconds), which is used as the zero-lag standard. Because of the different speeds of the two detectors, there will be a phase lag between the two signals, and to control it a phase shifter is introduced into one of the circuits. If both signals are given to an oscilloscope, a Lissajou figure will be observed (Fig. 2.8). By adjusting the phase shifter, a zero phase lag is obtained in which case the Lissajou figure transforms into a straight line. With knowledge of the experimentally observed phase angle and ω, the response time is obtained. This method has limitations; first, it gives only one predominant response time and does not give information about other time constants. Second, it does not gie information about the real form of the relaxation curve, such as whether it is exponential or not, and thus creates doubts about the presence of traps and related parameters. It is usual, therefore, to employ an oscilloscope to record the time evolution of the photovoltage. According to the response time of the photosensor, the most appropriate time base of the oscilloscope is selected. If the response is fast, several additional precautions are needed and the time constant of the circuit should be examined, as given by

$$\tau = (R_L + R_S)C_d \tag{2.20}$$

where R_L and R_S are load and series resistance, respectively, and C_d is the capacitance of the detector, which is proportional to the active area of the detector and inversely proportional to the square root of the applied bias voltage. The fastest response is obtained by using a small detector area and a low value of load resistance. Of course, this reduces the photoresponse, and a compromise is made between response time and responsivity. A typical circuit for measuring a fast response photosignal is shown in Fig. 2.9. In this case, reverse bias voltage is applied to the high-speed photodiode. R_V is a resistance that can be adjusted so that the voltage across the photodiode, and hence its capacitance can be controlled.

The above-mentioned method of transient studies of photosignals is good enough to observe the form of the relaxation curve and conclude whether it is exponential or deviates from it. As a result, the observed data are sufficiently accurate to plot a graph of log

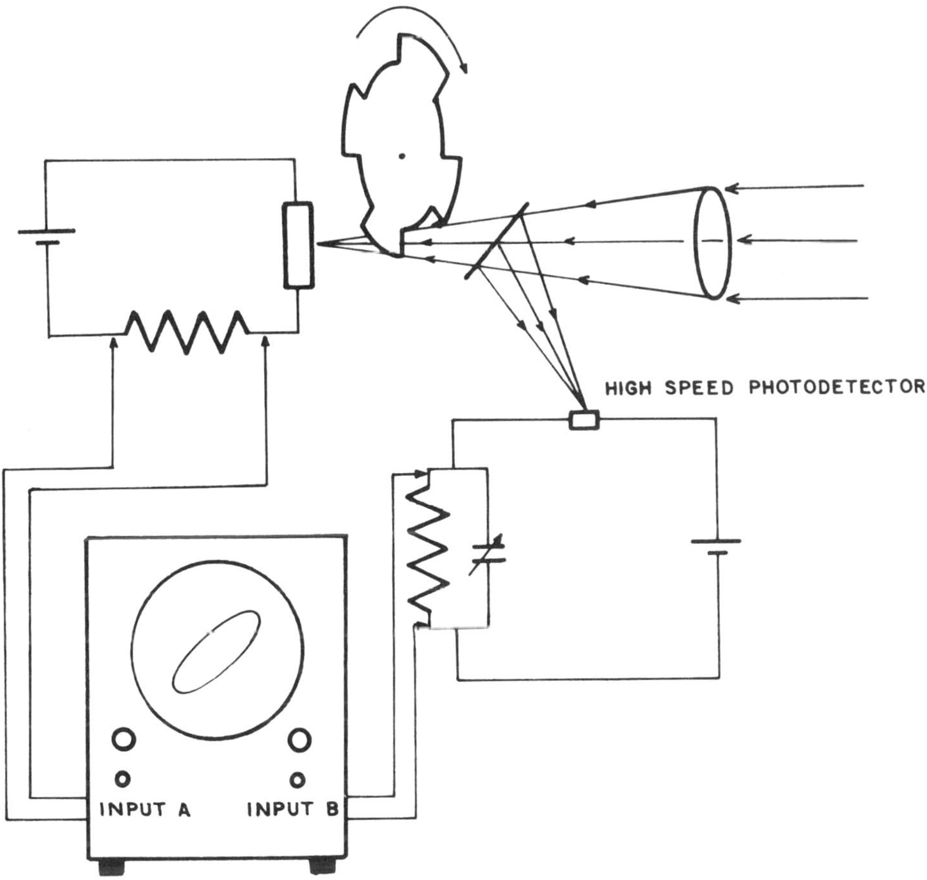

FIGURE 2.8 A basic circuit for determining $\tau_{response}$ by the phase-shift compensation technique.

photocurrent versus time and identify the presence of two or more time constants (not really, but two or three slopes; a different slope does not necessarily mean different time constnts [35]). In short, the measurement methods generally employed are just adequate for conventional types of analysis of the relaxation curves, but they are not suited for mathematical processing where digitized data are required.

Recently, the inadequacy of the method of analysis was stressed, and a new outlook was suggested by Joshi [35–37]. To improve our understanding of transient photoconductivity and separate the contribution from the delayed component, it is necessary to analyze

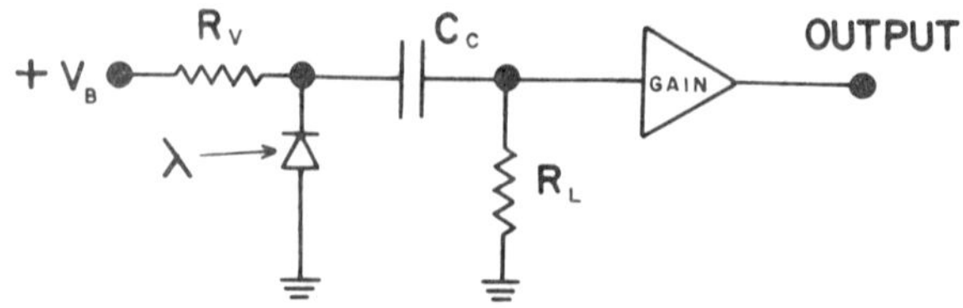

FIGURE 2.9 A conventional electrical circuit for high-speed photoresponse measurements.

the Fourier transform of the relaxation curve, and for this (and for other reasons) a large number of data points have to be recorded in digitized form (for details, see Chapter 3). For Fourier transform analysis of 2^n (n = 10 or 11 for better resolution, i.e., 1024 or 2048) data points are required.

Because of recent advance in electronics and microprocessors, modern equipment such as oscilloscopes, signal averagers, and electrometers are provided with digital outputs. The obtained transient response curve can be stored, and then any desired part of the curve can be enlarged on the time axis and details of it can be recorded. With the help of a standard interface (RS 232 or IEEE 488), it is possible to record a transient photocurrent in digitized form and use it for further data processing such as fast Fourier transforms and autocorrelation functions.

Fortunately, because of the availability of high-speed microprocessors, now it is possible to collect such data in a regime of nanoseconds or even picoseconds. A simple method is described below.

Photovoltage can be obtained from any of the above-mentioned methods and input to a digital high-speed voltmeter (or other similar instrument) to be converted into digitized form and recorded with the help of a high-speed computer. Microsecond response time can be recorded in 1024 data points of equally spaced intervals. Special equipment for high-speed data recording is also available commercially.

Having a large number of data points in digitized form, data handling and data processing become much easier. This stimulates the search for new techniques for data analysis. Joshi and Martin proposed a method for analyzing the existence of the delayed component with the use of the Fourier transform technique [36]. The details of this method of analysis will be discussed in Chapter 3.

Here it is worth mentioning that new techniques are being developed to analyze the combinations of two or more exponential curves. In fact, the lack of a proper method of analysis is a problem not only in photoconductivity or photoluminescence but also in many other branches of science such as nuclear physics and biophysics. Developments in measurement systems and hence in analysis and interpretation will have wider applications in the near future.

2.6. INSTRUMENTAL NOISE CONSIDERATIONS

Noise is a crucial parameter in measurement systems in general and in photoconductance in particular. Reduction in noise not only makes measurements more reliable and reproducible but also increases the limit of detection capability. Here, we will discuss noise aspects that are pertinent to experimental measurements; others will be discussed in Chapter 5.

For best performance, both the proper method and low-noise electronic equipment should be employed. From the point of view of weak signal detection, detector noise as well as amplifier and preamplifier noise should be as low as possible. The detectors used in the far-infrared region are cooled at liquid helium temperature. Typical examples are Insb, GaAs, and germanium photoconducting detectors [38—40]. These detectors at low temperature have high dynamic impedance and relatively large bandwidth. These features make it rather difficult to design preamplifiers having low noise over the entire detector bandwidth. This problem has been tackled by Brown [41], who proposed a design for an ultralow-noise, high-impedance preamplifier by using field effect transistors (FETs). The electronics of the design of low-noise amplifiers and preamplifiers is beyond the scope of this volume. There are several excellent books and articles available on the subject [e.g., 42,43], and readers are referred to them. However, in the experimental aspects of photodetection and measurement techniques, it is necessary to address amplifier noise and its effect on detector performance. It is obvious that an amplifier (independent of type) amplifies not only the photosignal but also the noise introduced at the input level. Moreover, it also creates extra noise, which varies from instrument to instrument. Hence, the S/N ratio at the output of the amplifier is different from the one at the input.

The role of the preamplifier can be understood with the help of Fig. 2.10. The magnitude of the excess noise depends upon the design of the preamplifier, matching impedance, and the type of noise at the entrance and its spectral distribution. The noise introduced by the amplifier is characterized by the so-called figure of noise defined as [44]

$$F = \frac{(S/N)_{input}}{(S/N)_{output}} \tag{2.21}$$

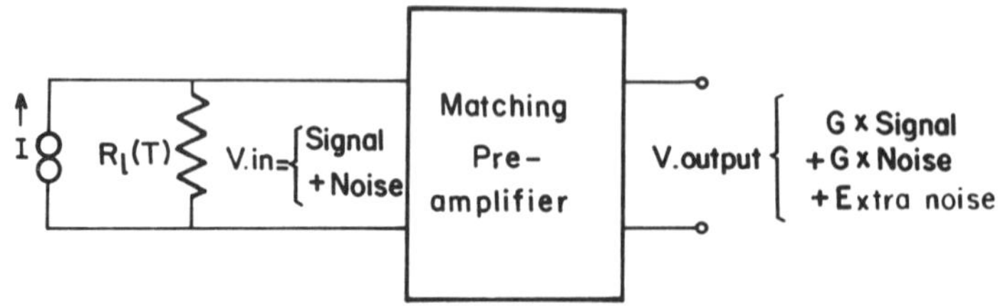

FIGURE 2.10 Amplifier contribution to noise.

Here N is the magnitude of noise generated at the source resistance R_L used in the biasing circuit. If necessary, the source resistor can be kept at a lower temperature to reduce the level of noise. Johnson noise at the source is kT df for bandwidth df, and the noise at the output is GkT df + extra noise. Similarly, the signal at the output is GS, where G is the gain of the instrument. Taking these values into account, the F factor can be given as [44]

$$F = 1 + (\text{extra noise})/GkT\ df \qquad (2.22)$$

The F value for a particular gain and for a given bandwidth is available for certain amplifiers, and hence the noise can be estimated. Here we have assumed that the dominating noise is Johnson noise; however, for other types of noise the F value can be obtained by using Eq. (2.21). Depending on the type of noise, S/N values for several photodetectors have been calculated and are available for ready use [44].

In addition to the above-mentioned noise, amplifiers create perturbations in the signal and mask or modify finer details of the information. This can be considerably reduced by adapting the detector and the preamplifier together. Electronic elements at the entrance of the preamplifier should be matched with respect to the resistance of the detector. Moreover, the preamplifier should not alter the speed of the photodetector. In some detectors, for example, indium antimonide, pertinent data are available (Section 1 of General Electric Bulletin LD-1). Recently, high-performance detectors have become commercially available with the preamplifiers.

Requirements of a preamplifier sometimes depend on the specific application. A typical example is the infrared photodetectors that are employed in space (say, in satellites). In this case, one has to consider the presence of spikes originating from the absorption of charged particles whose rate of flow and intensity are difficult to predict. It is therefore necessary to eliminate the effects of spikes with a special preamplifier. In a few cases preamplifier electronic circuits have been developed for very specific purposes [41, 45].

The experimental setup and preamplification system affect not only the magnitude of noise but also its frequency spectrum. Details will be discussed in Chapter 5.

2.7. METHODS FOR MODULATING LIGHT SIGNALS

As mentioned earlier, when the photosignal is weak, it can be separated from the dark current by modulating a light signal. This makes the signal alternating, and it can be amplified by a suitable ac or phase-sensitive amplifier. There are several methods for modulating a light beam. The commonly used ones are given below.

Figure 2.4 shows a chopper for modulating the signal; however, there are several techniques that are currently used for obtaining radiation pulses. The selection of the proper method depends on the requirements of the modulation frequency. For 5–3000 Hz a mechanical chopper seems to be adequate. For modulation frequencies higher than this, pulsating light sources such a stroboscopic lamps can be used.

The modern electronic stroboscope employs a gas-filled discharge tube to produce very short, repetitive, intense light flashes. A typical flash duration is about 1 μsec, and the flashing rate can be adjusted to between 100 and 150,000 per minute. In these cases, the dark interval is longer than the duration of the light, and hence the modulation is asymmetrical. Such light sources are really not convenient for the measurement of the spectral response of a photoconductor, particularly when the photoresponse is not fast.

In some special applications, the requirement of the modulation frequency is still higher, and then the electrooptical shutter (Kerr cell) is employed [46]. This device is able to interrupt a beam of linearly polarized light as much as 10^{10} times per second. The basic construction of this cell is given in Fig. 2.11. The system consists of a liquid cell of nitrobenzene fitted with two electrodes and placed between two crossed polarizers. The beam enters the cell in such a way that the direction of the polarization makes an angle of 45° with the direction of the applied field. The electric field applied to the electrodes decomposes the polarization plane of the beam into two components parallel and perpendicular to the electric field direction. Thus, the light beam emerging from the Kerr cell is circularly polarized because the two components travel with different velocities and have a phase difference. Circular polarization permits the transmission of radiation. In the absence of an electric field, no radiation is passed because the polarizer and analyzer are perpendicular to each other. In short, the effect is that of the electric field opening a gate for the light beam.

For certain purposes, particularly investigations in ultrahigh-speed photodetectors, pulses of still higher frequency are required,

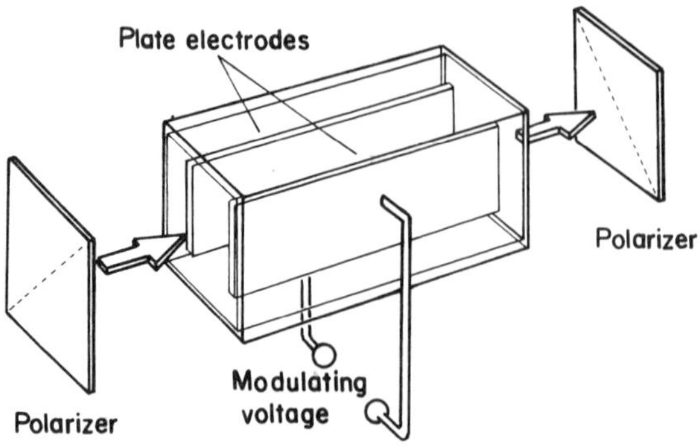

FIGURE 2.11 The design of a Kerr cell used for obtaining high-frequency pulses of radiation.

and this is now possible with laser diodes as a radiation source. With the help of these, the frequency limit could be extended further [47–49].

The above discussion might lead to the conclusion that a high chopping frequency is always desirable to detect weak signals—the higher the frequency, the better the photodetection. However, this is incorrect. There is a limit to increasing the frequency that is determined by the response time (not the lifetime). When the chopping frequency is high for a given rise time, then the time during which a photoconductor receives the radiation is not sufficient for the photoresponse to reach its saturation value.

Let us consider a conventional and widely used technique, namely, square modulation of radiation. Let t_{ill} be the duration of light shining on the sample. If this time is longer than the rise time, then the photocurrent reaches its saturation value (see Fig. 2.12a). Otherwise, the photocurrent starts to decay before it attains its steady state value. Naturally, the photosignal is reduced; the exact magnitude of the reduction depends on the ratio of rise to t_{ill} and also on the initial part of the rise curve. Obviously, if the chopping frequency is too high compared to the rise time, then the photosignal becomes too small to be detected.

In the case of asymmetrical square modulation (Fig. 2.12c), illumination time is not equal to the dark time ($t_{ill} \neq t_{dark}$). The forms of the rise and decay curves depend upon the ratio t_{ill}/t_{dark}; the limiting cases are discussed by Ryvkin [21]. Since the highest attainable value of the photocurrent depends on the duration of the illumination (t_{ill}), the photosignal varies with this duration. In general, measurements using asymmetrical square modulation do not provide any additional advantage over symmetrical modulation, and

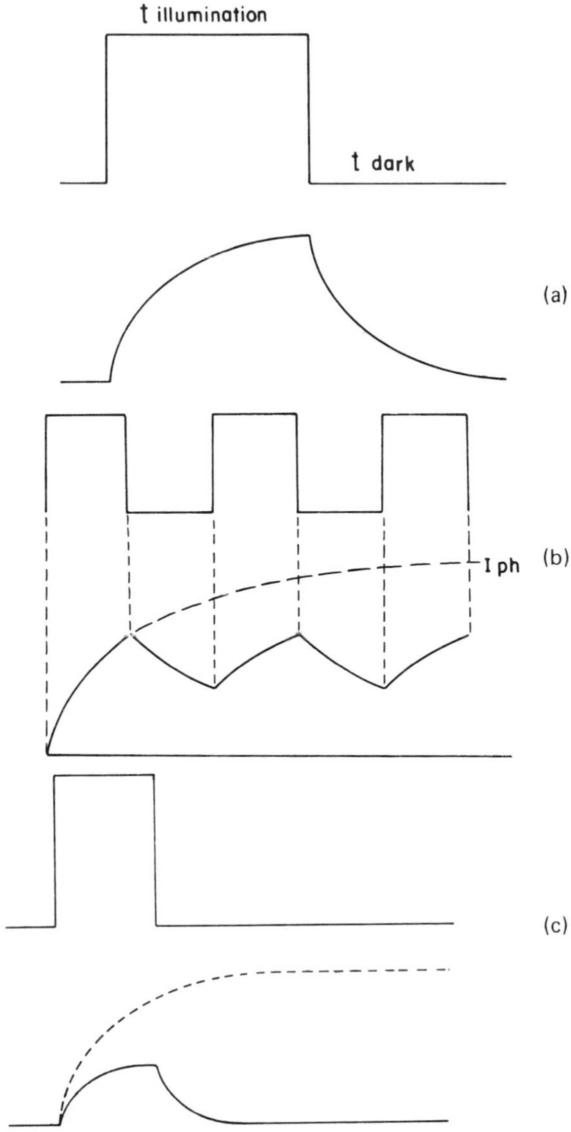

FIGURE 2.12 Effect of duration of light on the magnitude of the photosignal. (a) Duration of illumination equal to or greater than time required to attain steady state photocurrent; (b) duration of illumination less than time required to attain steady state photocurrent; (c) asymmetrical light pulse.

therefore the details are not discussed here. However, it is mentioned here because certain light sources, such as stroboscopic lamps, have asymmetrical pulses, and proper precautions should be taken to increase the response.

2.8. LIGHT SOURCES

For recording a photoconductivity spectrum, the light source should have a continuous spectrum in the region of interest and also should have sufficient power in that region that the detector can see it. The superposition of atomic spectral lines on a continuous spectrum is not desirable, but in some cases it is unavoidable. A monochromatic source such as a laser diode can be used when its wavelength matches the detector peak response. Such a source is very useful for optical communications and optical computers. For a spectral study one has to search for a proper radiation source. A few useful sources and their spectral distribution are given below. According to the convenience and the techniques of measurement, the spectral ranges are generally classified into the following groups

Far-ultraviolet region	125–300 nm
Ultraviolet region	300–400 nm
Visible region	400–700 nm
Near-infrared	700–1000 nm
Infrared	1–5 μm
Far-infrared	5 μm

The far-ultraviolet region comprises wavelengths between 125 and 300 nm, while the near-ultraviolet region covers wavelengths from 300 to 400 nm. Even though this separation is arbitrary, it helps to select a proper source. Certain lamps such as the high-pressure mercury lamp have an emission starting from 300 nm, the near-ultraviolet region, while phosphor-coated mercury lamps cover nearly the entire ultraviolet region.

2.8.1. Conventional Light Sources

Independent of spectral range, we classify light sources into two groups: conventional and nonconventional. The latter consists of solid state light-emitting and laser diodes. These are not widely used in laboratories, but I am sure that within a few years they will occupy an important place and therefore they deserve separate consideration here.

a. Arc Sources

The most useful and efficient sources in the ultraviolet region are arc sources [50–56; see also GE bulletins]. They are formed by passing a current between electrodes separated by gas or vapor. The passage of current causes atomic excitation in the medium, and upon deexcitation radiation is emitted. Depending upon the gas and its pressure, part of the emitted radiation is in the ultraviolet region. There are two types of arc lamps: open and closed. A typical example of the open lamp is a carbon arc [46] in which a discharge takes place in the atmosphere. A mercury vapor lamp is an example of a closed arc [51].

In the far-ultraviolet region (125 nm to about 390 nm), deuterium lamps [51] are more convenient. They are commercially available in both continuous mode and pulsed form. Photoconductivity in the far-ultraviolet region is not very frequently studied, because of both the limited applications and the unavailability of solid state photosensitive devices. Constant efforts are being made to search for high-band-gap photosensitive semiconductors to extend applications in this spectral region.

A deuterium with tungsten lamp also has spectral emission toward the infrared range and is suitable up to 1800 nm. The peak response is obtained near 1200 nm. Figure 2.13 shows a typical spectrum of a deuterium with tungsten lamp. The attractive feature of this lamp is that the response is smooth from 800 to 1800 nm.

There are several types of mercury lamps, and in all of them some percentage of the radiation is of wavelengths shorter than 280 nm. In many cases, unwanted UV radiation is filtered out. The spectral radiation distribution depends upon the pressure of mercury in the lamp. With low pressure, the continuous radiation is weak

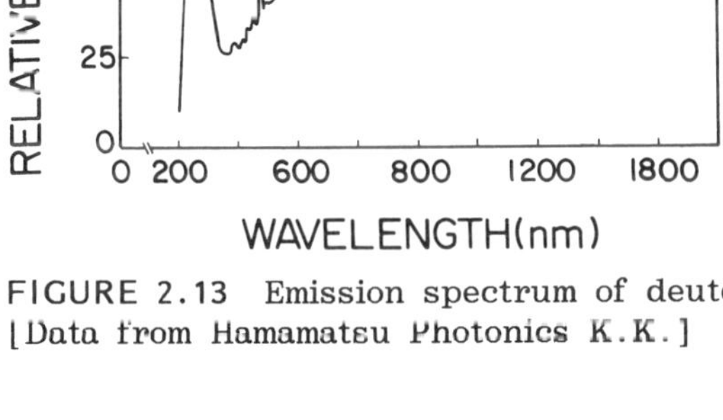

FIGURE 2.13 Emission spectrum of deuterium with tungsten lamp. [Data from Hamamatsu Photonics K.K.]

and the spectrum is dominated by strong atomic transitions. As the pressure increases, the intensity of the continuous radiation increases substantially. Figure 2.14 shows the spectral intensity per watt input for various pressures of mercury.

In the near-ultraviolet region, phosphor-coated mercury lamps are also used. These have a very narrow but useful spectral range (255–320 nm), which depends on the selected phosphorus material. One of the typical spectra is shown in Fig. 2.15. The spectral shape depends on the coating material.

The sunlamp is an ultraviolet lamp whose spectral range is restricted [52]. The Council of Physical Therapy of the American Medical Association defines a sunlamp as a lamp that, at a specified distance, emits an ultraviolet radiation not differing essentially from that of natural sunlight in the clearest weather at midday, in midsummer, at midlatitude and sea level, with total intensity and spectral wavelengths extending from 290 nm to 313 nm, and does not emit an appreciable amount of ultraviolet of wavelengths shorter than 280 nm. As mentioned earlier, all mercury lamps emit radiation wavelengths shorter than 280 nm. By filtering shorter wavelengths, a mercury lamp can be made to conform to the definition of a sunlamp.

A black light fluorescent lamp is, in fact, a low-pressure mercury lamp [50]. The inside wall of the lamp is coated with fluorescent or phosphorescent material that absorbs 2537-Å radiation and emits a broad band near 3600 Å. The glass used does not transmit wavelengths shorter than 2800 Å. Figure 2.16 shows the emission spectrum of cerium-activated calcium phosphate. This is a good source in terms of smooth continuous profile in the range

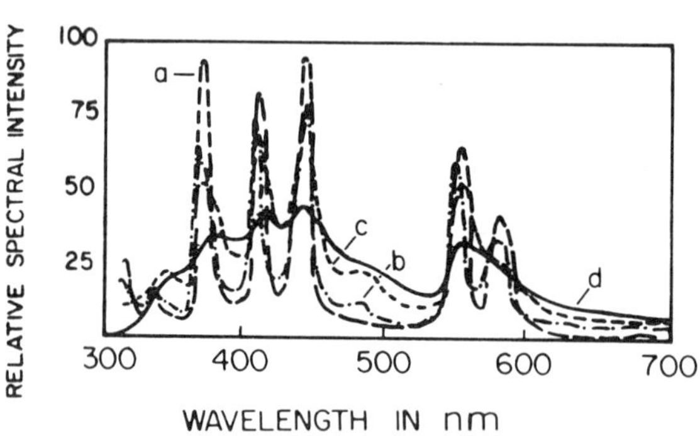

FIGURE 2.14 Variation of the emission spectra of a mercury lamp with pressure. a, 30 atm; b, 75 atm; c, 165 atm; d, 285 atm. [From Ref. 52.]

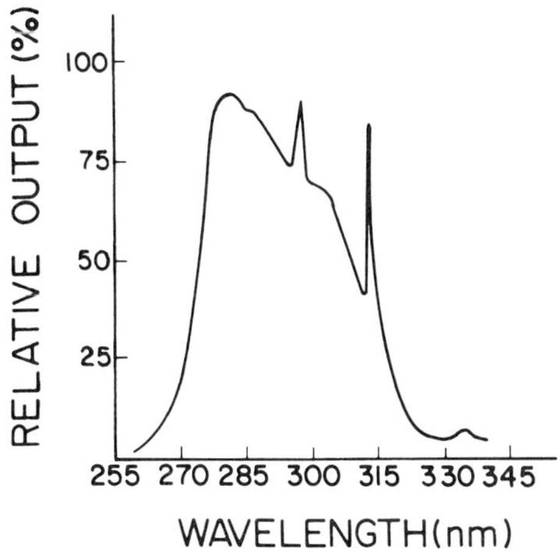

FIGURE 2.15 Spectral distribution of phosphor-coated mercury lamps. The spectral form, the intensity, and the position of the peak can be tailored by adjusting the phosphor material. [Data from Hamamatsu Photonics K.K.]

≃340–440 nm, because atomic lines, unlike the xenon spectrum, are absent and it has reasonably good output for recording the photoconductivity spectrum.

The zirconium arc [53] is probably a unique form of arc that approaches a point source. It is operated at high temperatures and hence emits ultraviolet radiation. When operated in atmosphere it emits strong radiation from 260 to 320 nm.

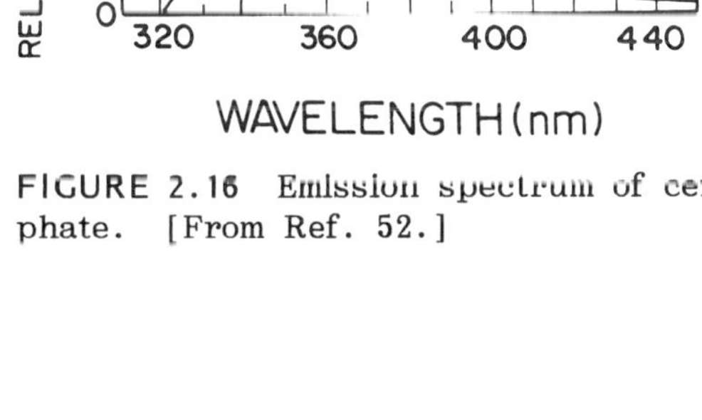

FIGURE 2.16 Emission spectrum of cerium-activated calcium phosphate. [From Ref. 52.]

Xenon arcs between tungsten electrodes and operating at high pressure are efficient sources of visible and ultraviolet radiation and cover the spectral range from 200 to 2000 nm [54]. However, the spectrum is superimposed with several atomic spectral lines, and therefore care should be taken to avoid the odd effects of these spectral lines in examining the photoconductivity spectrum.

Since 1960, broadband emission sources such as mercury-xenon arc lamps have been available (see General Electric bulletins). In this case 80% mercury and 20% xenon is used. A mixture of high-intensity ultraviolet emission together with the lines of xenon in the infrared region is observed. Thus the emission spectrum runs from the ultraviolet to the infrared. Such lamps are also available in continuous and pulse modes.

The carbon arc lamp is a good choice for the visible and ultraviolet region (≈ 350—700 nm). It is very commonly used for examining photoconductivity in wide-band-gap semiconductors, because of its high radiating flux and nearly uniform output in the region 350—700 nm. The spectrum of the carbon arc is continuous, superimposed by the atomic lines. The peak lies very close to the violet region (370—430 nm). The intensity of this peak increases very rapidly with the current of the arc. The overall intensity in the ultraviolet region is also increased remarkably.

One advantage of the carbon arc is that the intensity of the radiation in the ultraviolet region can be increased by using transition metal cored carbon [53]. The effects of the core metal, applied voltage, and current on the emission spectrum of the carbon arc has been carried out by the National Carbon Company, and the information is available from the company. A few typical spectra are given in Fig. 2.17.

Arc lamps are generally unstable; that is, the voltage and current passing through them fluctuate about their mean values. This instability stems from the fact that they have negative resistance. In designing the power supply, this aspect is taken into account, and the inherent instability is partially overcome by introducing a resistance (or reactance in ac arcs). This is called ballast resistance (or reactance), and with its help the dynamic characteristics are improved. A detailed study has been carried out by Copeland and Sparing [55].

There are additional factors such as temperature and pressure of the gas that also influence the instability, and therefore its complete elimination is very difficult. Nevertheless, in order to obtain the normalized spectral response curve it is necessary to know the output power precisely at the time the photoconductivity measurement is carried out. The presence of such fluctuations demands that the measurement of the photoresponse and the output power at the source be carried out simultaneously; this can be achieved by

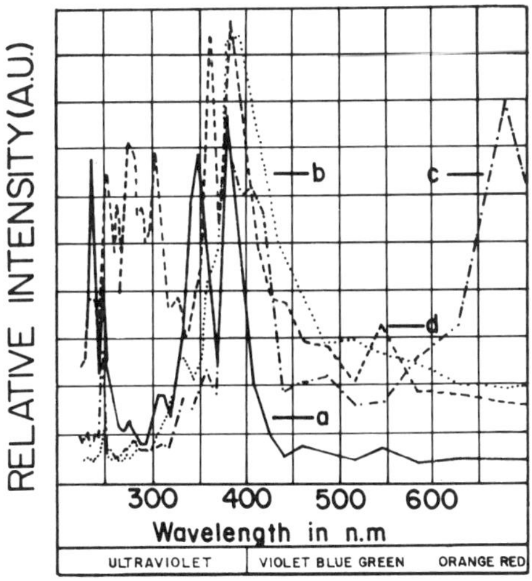

FIGURE 2.17 Spectral energy distribution of carbon arc. The emitted energy spectrum can be shifted toward the low-energy side by changing the core metal. (a) Cobalt core; (b) polymetallic core (80 A current); (c) strontium core; (d) polymetallic core (60 A current). [From Ref. 52.]

diverting a part of the incident beam, say 1%, to the thermocouple or to the calibrated power meter. When an incandescent radiation source is used, simultaneous measurements are not required, since the output power is nearly stable.

The list of sources given above is by no means complete, but it provides an idea of what kind of arc light sources could be used for recording photoconductivity spectra. The intensity of the radiation depends upon several factors such as pressure of the gas, distance between the electrodes, and applied voltage. Therefore, the intensity of this spectrum should not be compared directly with the intensity of any other spectra.

b. Incandescent Radiation Sources

An incandescent body is defined as one whose radiation is due to its temperature. An important characteristic feature of this type of radiation is its broad continuous spectrum. These sources radiate very little in the ultraviolet region; what they really emit is visible to infrared radiation. Thus, they are not only weak but also inefficient ultraviolet emitters.

The laws governing radiation emitted from incandescent bodies were established long ago both theoretically and experimentally, and our present level of knowledge permits us to calculate emitted energy with high precision. The total amount of radiated energy is proportional to the fourth power of the temperature, and for a blackbody the relation between temperature and energy is given by the Stefan-Boltzmann law [56] according to

$$W = \underline{\sigma} T^4 \tag{2.23}$$

where W is the energy radiated in watts per square centimeter of surface, T is the absolute temperature of the body, and $\underline{\sigma}$ is a constant equal to 5.67×10^{-12} W cm^{-2} K^{-4}.

It is clear from this equation that the total amount of energy radiated by an incandescent body increases extremely rapidly with the fourth power of temperature. The distribution of the energy is given by Planck's law and is expressed as [57]

$$f(\lambda T) = 2hc\lambda^{-5}/\exp(h\nu/KT) - 1 \tag{2.24}$$

where K is the Boltzmann constant.

Substituting the numerical constants for the 100-Å bandwidth, the energy distribution can be given as

$$E_\lambda \, d\lambda = \frac{3.7405 \times 10^{-12} \lambda^{-5}}{\exp(1.4388 \lambda/T) - 1} \; \text{W/cm}^2 \tag{2.25}$$

where E_λ is the energy in watts per square centimeter per 100-Å band of spectrum.

A blackbody is widely used as a radiation source in the near-infrared and middle infrared regions. One advantage of this source is the complete absence of atomic spectral lines. Figures 2.18 and 2.19 show the temperature derivative of photon emittance and integrated radiant photon emittance, respectively, for various temperatures.

To calculate important detector parameters such as quantum efficiency and detectivity, the photoconductivity spectrum must be normalized with respect to incident radiation measured in either watts or photons. Figures 2.18 and 2.19 could be very useful for this purpose. In addition to this, Tables 2.1 and 2.2 give the power radiated by the blackbody source at various temperatures. With the help of these tables, energy or photon distributions over the wavelength range 1–30 μm can be estimated. The values given in these tables represent emission into a complete hemisphere in front of the blackbody source. To obtain emission per unit solid angle normal to the source, these values should be divided by π.

Incandescent lamps are typically characterized by their color temperature, life, and spectral output, and uniformity, size, and shape

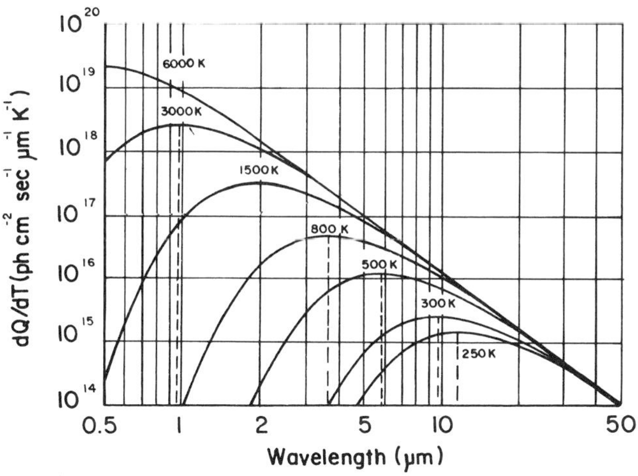

FIGURE 2.18 Temperature derivative of photon emittance. [From Photonic Handbook, 1985, courtesy of Photonic Spectra Magazine.]

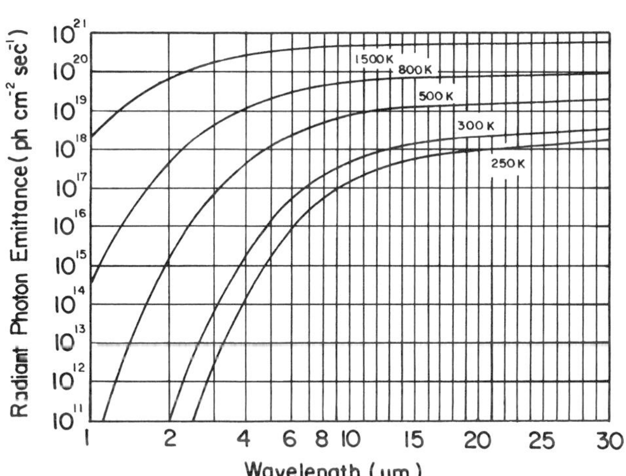

FIGURE 2.19 Integrated radiant photon emittance as a function of wavelength. [From Photonic Handbook, 1985, courtesy of Photonic Spectra Magazine.]

TABLE 2.1 Total Number of Photons and Fraction over the Waveband 0 to $-\lambda$ Emitted per Square Centimeter per Second by Blackbodies at Different Temperatures and Wavelengths

Temp. °C	Total number of Photons	Wavelength										
		1 μm	2 μm	3 μm	4 μm	6 μm	8 μm	10 μm	15 μm	20 μm	25 μm	30 μm
600	1.04×10^{20}	9.8×10^{-6}	9.9×10^{-3}	7.6×10^{-2}	1.89×10^{-1}	4.15×10^{-1}	5.8×10^{-1}	6.87×10^{1}	8.31×10^{-1}	8.95×10^{-1}	9.29×10^{-1}	9.44×10^{-1}
800	1.92×10^{20}	1.4×10^{-4}	3.2×10^{-2}	1.5×10^{-1}	2.98×10^{-1}	5.34×10^{-1}	6.79×10^{-1}	7.68×10^{-1}	8.79×10^{-1}	9.26×10^{-1}	9.51×10^{-1}	9.65×10^{-1}
1000	3.21×10^{20}	8.45×10^{-4}	6.8×10^{-2}	2.34×10^{-1}	3.99×10^{-1}	6.24×10^{-1}	7.49×10^{-1}	8.22×10^{-1}	9.10×10^{-1}	9.46×10^{-1}	9.64×10^{-1}	
1500	8.68×10^{20}	1.09×10^{-2}	1.96×10^{-1}	4.22×10^{-1}	6.85×10^{-1}	7.64×10^{-1}	8.5×10^{-1}	8.98×10^{-1}	9.5×10^{-1}	9.71×10^{-1}		
2000	1.83×10^{21}	4.20×10^{-2}	3.31×10^{-1}	5.65×10^{-1}	7.04×10^{-1}	8.4×10^{-1}	9.02×10^{-1}	9.34×10^{-1}	9.68×10^{-1}			
3000	5.42×10^{21}	1.58×10^{-1}	5.42×10^{-1}	7.34×10^{-1}	8.30×10^{-1}	9.13×10^{-1}	9.48×10^{-1}	9.66×10^{-1}				

Example: Total number of photons at 3000°C is 5.42×10^{21}. The number of photons between 0 and 3 μm at 3000°C is $5.42\times10^{21}\times7.34\times10^{-1}$.

TABLE 2.2 Power (W/cm^2) Emitted by Blackbody at Various Temperatures

Temp. (°C)	Total	1μm	2μm	3μm	4μm	6μm	8μm	10μm	15μm	20μm
600	3.3	6.5×10^{-5}	3.48×10^{-2}	1.9×10^{-1}	3.87×10^{-1}	6.66×10^{-1}	8.11×10^{-1}	8.84×10^{-1}	9.57×10^{-1}	9.8×10^{-1}
800	7.50	7.65×10^{-4}	9.40×10^{-2}	3.29×10^{-1}	5.37×10^{-1}	7.75×10^{-1}	8.80×10^{-1}	9.29×10^{-1}	9.75×10^{-1}	
1000	1.49×10^{1}	3.85×10^{-3}	1.76×10^{-1}	4.54×10^{-1}	6.50×10^{-1}	8.43×10^{-1}	9.19×10^{-1}	9.53×10^{-1}		
1500	5.6×10^{1}	3.30×10^{-2}	3.99×10^{-1}	6.76×10^{-1}	8.16×10^{-1}	9.27×10^{-1}	9.65×10^{-1}			
2000	1.51×10^{2}	1.18×10^{-1}	5.77×10^{-1}	8.00×10^{-1}	8.94×10^{-1}	9.61×10^{-1}				
3000	6.45×10^{2}	3.40×10^{-1}	7.82×10^{-1}	9.11×10^{-1}	9.56×10^{-1}					

The table contains the total and the fraction of the power over wavelengths 0 to λ.
Example: The total power emittited at 300° K is 6.45×10^{-2} W. The power emitted between 0 to 3μm is $6.45\times10^{-2}\times9.11\times10^{-1}$ W.

of the filament (for focusing on the slit of the monochromator). The color temperature varies from 2100 to 3300 K, as the wattage of the bulb varies from 0.5 to 35 W. This gives only a rough idea of the color temperature of a source (which is not the temperature of the source).

An ordinary tungsten filament lamp is a typical example of this type of source. It consists of a coiled filament of fine tungsten wire located in a gas bulb that is highly evacuated or filled with an inert gas. In both types, light output is achieved through the incandescence of a tungsten filament. The size and shape of the filament are varied according to the purpose. The radiation emitted is mainly in the visible and near-infrared regions. The total radiated power increases sharply with temperature, and therefore a tungsten lamp is operated at the highest possible temperature. The light output decreases over the life of the lamp due to evaporation of tungsten and its deposition on the interior of the glass surface. In the case of gas-filled lamps, tungsten evaporation is partially suppressed by the gas. Its output power is relatively increased. Gas-filled lamps have a longer life. These lamps are available in several sizes, and voltage and current requirements can be designed according to the output power desired. They are easy to operate, widely available, and cover a reasonable spectral width (400—1800 nm depending upon the type of glass or quartz), and above all they are economical. These features make them popular for laboratory use.

The tungsten halogen lamp is an improved version of the tungsten lamp. In addition to an inert gas, it contains a suitable halogen compound. The inert gas suppresses the evaporation of tungsten, and the halogen compound reacts with the evaporated tungsten, preventing its deposition on the glass wall. A regenerative cycle redeposits tungsten on the filament and extends the life of the bulb.

Halogen lamps are recommended when higher luminous efficiency, longer life, and wider spectral range are required.

The color temperature of a lamp and its efficiency are related. The higher the temperature, the higher the efficiency; at about 3400 K the efficiency of a halogen lamp becomes close to 100%. For tungsten lamps, the relation is shown in Fig. 2.20. The spectral output of a tungsten lamp is in the range 280—2500 nm, and it is a good choice for recording photoconductivity spectra.

For the middle infrared region, many of the above-mentioned sources are not adequate. Generally, a blackbody source, Nernst filament, [49,58,59] and globar lamps [60] are employed. The spectral distribution of a blackbody radiator at a particular temperature T can be estimated with the help of Eq. (2.24) or (2.25) and Table 2.2.

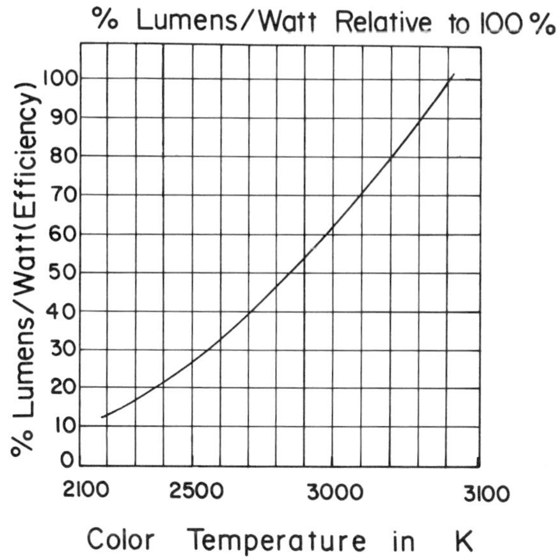

FIGURE 2.20 Typical relation between color temperature and efficiency for a tungsten lamp. The relation varies with a thorium coated tungsten lamp.

The globar source consists of a rod of silicon carbide a few centimeters long and about 4–6 mm in diameter. It is heated electrically to a temperature of not more than 1400 K. It has a reasonably uniform power distribution from about 1.5 to 15 μm. In this range it emits approximately 80% as much radiation as a blackbody. For longer wavelengths, the spectral output reduces considerably, but reasonable power can be obtained up to 40 μm.

The other commonly used source in the infrared region is the Nernst filament, which is made from zirconium and yttrium oxides. Platinum wires are connected at both ends as an electrode. At room temperature, the Nernst filament is a bad conductor, and therefore it should be heated externally by some suitable system until it becomes conducting. It used to be heated by flame, but now it is heated with a small furnace that houses the filament. The temperature of the furnace is controlled with a special circuit that is disconnected when the filament is sufficiently heated and becomes conducting. The spectral distribution of the filament is similar to that of a globar source in the middle infrared region [46,49]. The Nernst filament is an excellent source of radiation because its emission corresponds very closely to that of a blackbody radiator.

In the far-infrared region, experimental techniques are quite different. Neon and mercury arc sources are commonly employed.

Grating spectrometers are replaced by the fast Fourier transform spectrometer. The experimental details are discussed in Chapter 7.

2.8.2. Nonconventional Light Sources

a. *Light-Emitting Diodes (LEDs)*

Light-emitting diodes (LEDs) are not considered light sources for photoconductivity spectral studies. However, these sources are useful for specific purposes such as examining the detector response for a particular wavelength (peak response). As early as 1965, it was observed that GaAs np junctions convert electrical energy into radiation with high efficiency (about 40%). As a result, considerable efforts were made to improve the techniques and search for new materials, and now there are several LEDs whose peak response varies from the visible to the infrared region. The development, design, construction, and electrooptical properties of LEDs is beyond the scope of this book. Therefore, no attempt is made to cover the details of these radiation sources. Here, we comment briefly on the spectral distribution of the output radiation.

The spectral response of an LED can be controlled by selecting material of the proper band gap. A slight change in the spectral position can be achieved by varying the concentration of the element in the alloy. It is also possible to shift the spectral position by introducing the proper impurity [61,62]. A typical example is $GaAs_xP_{1-x}$ alloy [61]. By changing the percentage of phosphorus and controlling the nitrogen impurity, the spectral response can be varied from red to yellow as shown in Fig. 2.21. Table 2.3 lists the composition and wavelength for which the peak response is reported for some highly efficient LEDs. It is worth mentioning that laser diodes are now commercially available and are quickly replacing LEDs.

b. *Laser Diodes*

In the mid-1980s we have witnessed an explosive growth of technology of semiconductor laser diodes [63]. Now high-speed, low-threshold current and long-life laser diodes are available at relatively low cost. Although they are being used as a radiation source in laboratories, they have specific applications. Since they are a monochromatic source of radiation, they cannot be used for the spectral study of photodetectors.

The best studied commercially available laser diodes are made from $Al_xGa_{1-x}As$—GaAs heterojunctions, and the output power is in the near-infrared region and dependent on the concentration of aluminum. The other widely used laser diode is made with a GaInAsP/InP structure. The type of diode selected depends on the

TABLE 2.3 Peak Response Energies of Selected Light-Emitting Diodes[a]

Alloy	Peak response (eV)
InGaP	2.2
GaAsP	2.0
InAlP	1.65
AlGaAs	1.41–1.51
GaAs	1.32
InGaAs	1.16
InGaAsP	0.95

[a]The energies given here are approximate and depend upon composition of the alloy.

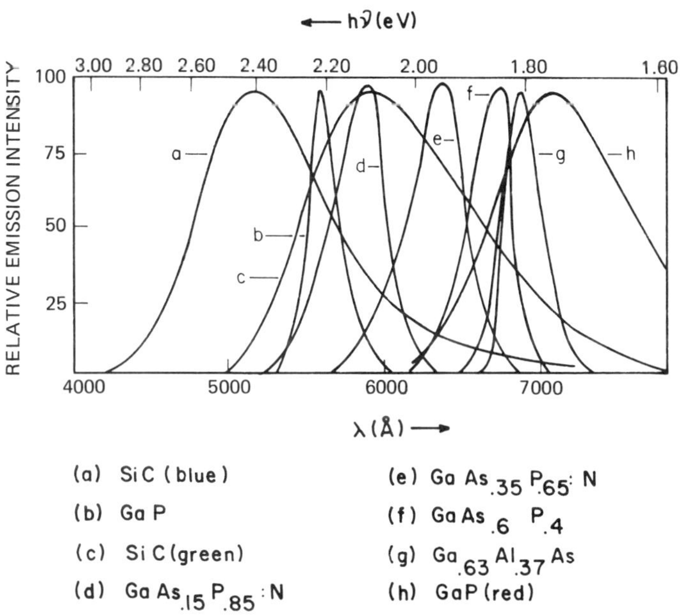

FIGURE 2.21 The emission spectra of some commercially available photodiodes. (a) SiC (blue); (b) Ga P; (c) SiC (green); (d) Ga $As_{.15}P_{.85}$:N; (e) Ga $As_{.35}P_{.65}$:N; (f), Ga $As_{.6}P_{.4}$; (g) $Ga_{.63}Al_{.37}As$; (h) GaP (red). The intensities are normalized to their peak values and should not be compared with each other. The positions of the maxima and the sensitivity vary with the composition of the alloy, design and the processing technology of the devices.

spectral range and applications. The concentration of the alloy gives flexibility in the adjustment of peak emission wavelength according to the detectivity of the photodetector.

The list of laser diodes is increasing very fast and would cover a considerable portion of the red and infrared spectral range. For a monochromatic source, laser diodes are ideal and are now also becoming affordable. Moreover, they have several applications. In the near future, they will play a dominant role as light sources.

REFERENCES

1. H. K. Henisch, Semiconductor Contacts: An Approach to Ideas and Models. Oxford Univ. Press, 1984.
2. R. Holmes, Electric Contacts, 4th ed., Springer-Verlag, New York, 1967.
3. F. A. Padovani, The Voltage-Current Characteristics of Metal-Semiconductor Contacts. In: Semiconductors and Semimetals (R. K. Willardson and A. C. Beer, Eds.), Vol. 7. Academic, New York, 1971.
4. J. G. Simmones, J. Phys. Chem. Solids 32,2581 (1971).
5. B. Schwartz (Ed.), Ohmic Contacts to Semicondcutors. Electrochemical Society, Inc., New York, 1969.
6. G. Weimann and W. Schlapp, Phys. Stat. Solidi 50, K219 (1978).
7. T. F. Connolly (Ed.), Solid State Physics Literature Guide, Vol. 4, Plenum, New York, 1972.
8. R. A. Ginley and D. D. L. Chung, Solid State Electron. 27, 146 (1984).
9. T. Takahashi and A. Ebina, Appl. Surface Sci. 11/12, 268 (1982).
10. W. E. Spicer, I. Lindau, P. E. Gregory, C. M. Garne, P. Pinetta, and P. W. Chye, J. Vac. Sci. Technol. 13, 780 (1986).
11. L. Ley, R. A. Pollak, F. R. McFeely, S. P. Kowalczyk, and D. A. Shirley, Phys. Rev. B9,600 (1974).
12. G. J. Lapeyre and J. Anderson, Phys. Rev. Lett. 35,117 (1975).
13. P. Masri and M. Lannoo, Surface Sci. 52,377 (1975).
14. J. D. Levine and P. Mark, Phys. Rev.B 182,926 (1969).
15. T. Murotani, K. Fujiwara, and M. Nishijima, Phys. Rev. 12B, 2424 (1975).
16. J. A. Nicholson, J. D. Riely, R. C. G. Leckey, J. G. Jenkin, and J. Liese-gang, Phys. Rev.B 18,2561 (1978).
17. M. McCrorey, M. Bujatti, and F. Marcelja, J. Phys. C9,4281 (1976).
18. J. Martin, R. Casanova, and N. V. Joshi, Phys. Rev. 36B,9703 (1987).

19. M. Ito and O. Wada, IEEE J. Quantum Electron. QE22,1073 (1986).
20. A. A. Turnbull, Br. J. Appl. Phys. 15,1051 (1964).
21. S. M. Ryvkin, Photoelectric Effects in Semicondcutors, Consultants Bureau, New York, 1964.
22. Y. Marfaing, Photoconductivity and Photoelectric Effects. In: Handbook on Semicondcutors, Vol. 2, North-Holland, Amsterdam, 1980, p. 417.
23. W. Wang, Solid State Electronics, McGraw-Hill, New York, 1966.
24. J. Tauc, Photo and Thermoelectric Effects in Semiconductors. Pergamon, Oxford, 1964.
25. H. Heyszenau, Solid State Commun. 35,561 (1980).
26. E. M. Uyukin, Sov. Phys. Solid Stat. 21,910 (1979).
27. T. S. Moss, Semiconductors and Semimetals, Vol. 2 (R. K. Willardson and A. C. Beer, Eds.). Academic, New York, 1966.
28. A. Ambroziak, Semiconductor Photoelectric Devices. Iliffe Books, London, 1968.
29. N. V. Joshi, L. Martinez, and R. Echeverria, J. Phys. Chem. Solids 42,281 (1981).
30. N. V. Joshi and H. Aguilar, J. Phys. Chem. Solids 43,797 (1982).
31. N. V. Joshi and N. Mireya Castillo, Chem. Phys. Lett. 41,490 (1976).
32. F. Gutman and L. E. Lyons, Organic Semiconductors. Wiley, New York, 1967.
33. P. Lenczewski and E. Fortin, Phys. Stat. Solidi (a)3,K267 (1970).
34. D. Barreto, J. Luengo, Y. De Vita, and N. V. Joshi, Appl. Opt. 26,5280 (1987).
35. N. V. Joshi, Phys. Rev. B27,6272 (1983).
36. N. V. Joshi and J. M. Martin, Phys. Lett. 113A,318 (1985).
37. N. V. Joshi and S. Swanson, J. Solid State Electron. Special Issue on Electro-Optics 30,105 (1987).
38. J. N. Crouch, Electro-Technology 75,96 (1965).
39. S. W. Stephens, Infrared Phys. 8,185 (1968).
40. D. B. Williams, Infrared Phys. 5,57 (1965).
41. E. R. Brown, Electron. Lett. 21,417 (1985).
42. F. N. H. Robinson, Noise and Fluctuations in Electronic Devices and Circuits. Clarendon Press, Oxford, 1974.
43. R. Bernoit and G. Bertolini, in: Low Noise Electronics in Semiconductor Detectors. (Bertolini and A. Coche, Eds.). American Elsvier, New York, p. 201.
44. R. H. Kingston, Detection of Optical and Infrared Radiation. Springer-Verlag, Berlin, 1978.

45. J. R. Houck and D. A. Briotta, Jr., Infrared Phys. 22,215 (1982).
46. E. Hecht and A. Zajac, Optics. Addison-Wesley, Reading, Mass., 1973.
47. N. V. Joshi, Unpublished report, "Monolithic Integration of Ultra-High Speed Photonic Devices." Physical Sciences Center, Honeywell, Minnesota.
48. K. Y. Lau and A. Yariv, IEEE J. Quantum Electron. 21,121 (1985).
49. R. S. Tucker, IEEE Trans. Electron. Devices ED32,2572 (1985).
50. W. E. Forsythe (Ed.), Measurements of Radiant Energy. McGraw-Hill, New York, 1937, Chs. 1 and 2.
51. A. E. Green (Ed.), Middle-Ultraviolet: Its Science and Technology, Wiley, New York, 1967.
52. L. R. Koller, Ultraviolet Radiation. Wiley, New York, 1965.
53. W. E. Forsyth and E. A. Adams, Fluorescent and Other Gaseous Discharge; Lamps, Murray-Hill Technical Div., New York, 1948.
54. General Electric Co. Lamp Bulletin LD-1.
55. P. Copeland and W. H. Sparing, J. Appl. Phys. 74,1524 (1948).
56. R. B. Leighton, Principles of Modern Physics. McGraw-Hill, New York, 1959.
57. J. M. Stone, Radiation and Optics. McGraw-Hill, New York, 1963.
58. W. Nernst, Z. Elektrochem. 6,41 (1899).
59. W. Nernst and W. Wild, Z. Elektrochem. 7,373 (1900).
60. V. Z. Williams, Rev. Sci. Instrum. 19,135 (1948).
61. E. W. Williams and R. Hall, Luminescence and The Light Emitting Diodes, Pergamon, Oxford, 1979.
62. R. N. Bhargava, Current status of visible LED. IEE Trans. Electron Devices ED-22,691 (1975).

3

Transient Photoconductivity

3.1. GENERAL INFORMATION AND IMPEDIMENTS TO ANALYSIS

The time-dependent behavior of the photocurrent is known as transient photoconductivity. As explained in Chapter 2, transient forms are generally recorded on an oscilloscope and, more recently but not frequently, they are recorded in digitized form with the help of a computer. Commonly observed forms of relaxation (transient) curves are shown in Fig. 3.1. Even in correctly performed experiments, several forms of rise and decay curves are reported. Photoresponse may vary widely with time, and, depending upon the presence of defect states (traps) and their distribution, the observed curve may also deviate frequently from a simple exponential. In principle, photoconductivity may also become one of the powerful techniques for studying optically active traps since the transient curves contain a wealth of information about the kinetics of charge carriers. Such information, however, is very rarely obtained in a true sense, simply because these curves are not easy to interpret. There are diverse impediments to analyzing such curves.

1. Several of the parameters, which are difficult to evaluate by other independent methods (density and distribution of optical active traps, their capture cross section, etc.), are involved in the kinetics of the charge carriers.

2. These parameters vary from sample to sample and are very sensitive to such factors as temperature, level of radiation, effects of contacts, and surface conditions.

3. The effects of several relaxation mechanisms are combined in the apparent time constant. The argument of the exponential might

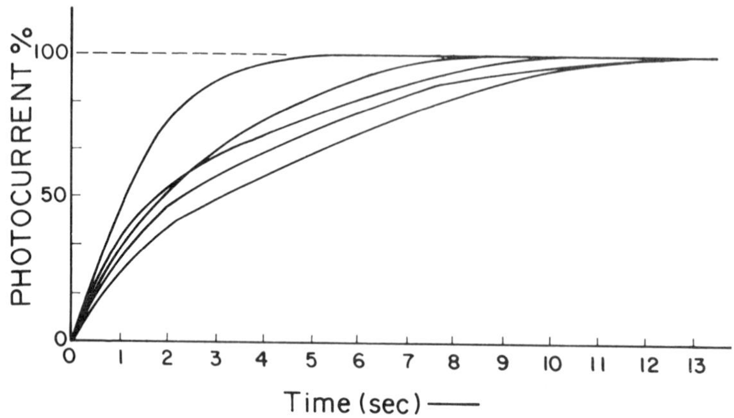

FIGURE 3.1 A few typical forms of relaxation curves. Parameters that influence the shape are discussed in the text.

be the reciprocal of the sum of the reciprocals of related time constants such as the bulk and surface time constants. When bulk and surface recombination rates are of the same order, then the contributions of both are reflected in the resultant relaxation time constant, which is given by [1]

$$\frac{1}{\tau_{eff}} = \frac{1}{\tau_{surf}} + \frac{1}{\tau_{bulk}} \tag{3.1}$$

where τ_{surf} and τ_{bulk} are the surface and bulk recombination time constants, respectively. In that case, there does not exist any mathematical method for separating them. The analysis of multicomponent exponential curves is a widely discussed problem [2–4], since it appears in a variety of experimental situations, and several techniques have been suggested for analysis [3–5], sometimes with only limited success [6,7].

There are other situations in which the observed decay time is not a single time constant but the sum of the reciprocals of the various time constants, as represented by Eq. (3.1). This can happen even in the absence of optically active traps. When the rate equation includes the contribution of the diffusion process, which is a second-order partial differential equation, then the solution of the equation (we will see the origin of these equations shortly) is represented by a Fourier series [8] whose coefficients correspond to different time constants τ_i. Generally, the first few terms could make a significant contribution, and in this case the observed decay time is written as [8]

$$\frac{1}{\tau_{obs}} = \sum_i \frac{1}{\tau_i} + cD \qquad (3.2)$$

where D is the diffusion coefficient and c_i can be considered a constant.

When a correctly carried out experiment confirms the presence of only one decay time constant, it means that the contribution of the second component is small and the surface recombination effect is negligible. This situation can be, but is not necessarily, achieved by preparing the surface carefully with a proper etchant and illuminating the sample uniformly.

4. Several models explain the observed form of the relaxation curves (within limited accuracy) [9,10]. The problem is not to search for the model but to remove the ambiguity, as there are a few models (or a few sets of parameters) that can explain the same trend of the relaxation curves [11].

5. Kinetics of the charge carriers leads to a set of nonlinear differential equations, and it is known that there is no closed solution for such a system. Moreover, until recently it was also thought that the search for a solution to the system of differential equations was an unrewarding task [11], and so sufficient efforts in that direction were not carried out.

6. Charge carriers are trapped in centers and reejected in the conduction or valence band. This means that a delayed contribution in the relaxation curve is expected; however, mathematical analysis of the delayed equation is not at such a level that it can be used directly for the present problem. Very recently [12], an attempt to detect and separate the delayed component has been reported. However, systematic work in this direction is needed.

7. The decay processes of excess charge carriers consist of several simultaneous and competitive mechanisms. Each of them is complex, and there are several variations in the process. For example, recombination can be linear or quadratic. It might take place directly or through the recombination centers. Recombination can be partly radiative and partly nonradiative. Due to these possible complications, it is not possible to estimate the contribution separately with the corresponding time constants.

8. The transient curve is very seriously influenced by the contact properties [9,13]. It is customary (but sometimes incorrect), to ignore the contact effects. When injecting contacts are present, the properties of the semiconductor under them are modified because

a. Carrier concentrations are changed.
b. Space charges are created.
c. Diffusion currents are set up.

In these circumstances, the analysis is practically impossible. To an experimentalist, the contacts and surface conditions must be improved to avoid the side effects.

9. A large amount of experimental data show that time-dependent photoconductivity is observed on the oscilloscope or on the recorders. The curve recorded in this manner shows only the general trend and not the detailed behavior. Quantitative analysis, naturally, requires a recording of such curves in digitized form [14]. This is now possible because of the availability of high-speed computers that enable experimenters to collect precise data quickly. Computers are also a help in data handling and statistical processing.

Every aspect mentioned above is an impediment to interpreting the time evolution of the photocurrent. It is not that those aspects cannot be handled, but a separate study for each of them is needed to understand them before analyzing the relaxation curves. For example, Schultz [15] has separated the bulk recombination contribution from surface recombination by measuring the amplitude and phase photoconductance in different ambient conditions. The theory developed for this purpose is based on the assumption that n = p and requires data from several values of surface effects have been detected separately and the contribution of recombination rates. In this way the volume effect could be examined. Obviously, the method proposed by Schultz cannot be generalized, because in a majority of cases, photoexcited charge carriers are stored in the traps or recombination centers for an appreciable time and the required condition, namely n = p, is not satisfied. Therefore the applicability of this method is limited. Similarly, the rate of emission of charge carriers from impurity centers has been investigated independently to examine the slow photoconductivity kinetics [16]. Such knowledge, is sometimes indispensable for evaluating the earlier models or understanding the recombination process.

The complexity mentioned here encourages a search for an alternative approach, namely, numerical methods. Recently, some investigators [17] solved the set of simultaneous nonlinear equations with a numerical technique using the Runge-Kutta sixth-order predictor method. The results were satisfactory, but only for a given set of parameters. The random walk approach was also successfully used [18]. Numerical approaches certainly explain a particular form of the relaxation curve in specific conditions; however, they cannot form a theoretical basis for an understanding of the kinetics of charge carriers and are not appropriate for predicting time-dependent behavior, even though the important trap parameters are known. This is understandable because the initial conditions such as density of charge carriers and the numbers of empty and filled traps are not kown precisely, and an error in them gives a solution considerably different from the real one. In short, all approaches of numerical methods are very sensitive to the initial conditions, which must be at least approximated—a difficult task.

I do not mean to suggest that such approaches do not provide useful information for a general understanding of the basic process. Significant information has been reported by Popescu and Henisch [19], who solved continuity equations (we will examine these equations later) by numerical methods. They obtained the carrier concentrations and the recombination rates as a function of the distance from the injecting boundary. The behavior of the recombination front and related parameters can be understood with the numerical solution. In fact, such information should be incorporated in the formal theory to explain the time evolution of the photocurrent.

Here we have mentioned some of the factors that clearly explain why the meaningful analysis of relaxation curves is still a major issue. This is not to say that every observed relaxation curve is influenced directly or indirectly by each and every factor, but great care must be taken in performing the experiment and interpreting the observed results.

3.2. TIME AND SPACE DEPENDENCE OF CHARGE CARRIERS: SIMPLE CASE (TRAP-FREE CASE)

The literature on photoconductivity relaxation curves is abundant [e.g., 9–11], and several attempts have been reported to explain the variety of experimentally observed curves. However, we start the discussion with a simple case. Traps always exist even in high-purity and in high-quality semiconductors. The trap-free case is referred to a case where the number of electrons being perturbed by traps is much less than the number of photoexcited electrons. Such situations are very frequently encountered.

Let us consider that $\underline{n}$ and $\underline{p}$ are the numbers of electrons and holes, respectively, created in the conduction and valence bands by optical excitation. Then the steady-state photoconductivity is given by [20]

$$I_{pho} = e(\mu_n \underline{n} + \mu_p \underline{p}) \tag{3.3a}$$

Since

$$\underline{n} = \alpha\eta \underline{I}_0 \tau_n \quad \text{and} \quad \underline{p} = \alpha\eta \underline{I}_0 \tau_p$$

the photoconductivity is expressed as

$$I_{pho} = e\ \alpha\eta \underline{I}_0(\mu_n \tau_n + \mu_p \tau_p) \tag{3.3b}$$

Neither $\underline{n}$ nor $\underline{p}$ is constant in the entire sample, as the generation rate varies with the depth from the illuminated surface according to Lambert's law of absorption. This causes a diffusion process into the bulk.

As we have seen earlier, it takes a certain time to attain steady-state photoconductivity. This is mainly because certain time is required to balance the generation and recombination rates for the electrons and holes. In such circumstances, the space and time variation of charge carriers can be examined by considering their flow, generation, and recombination rates in a small volume dv [21]. Let τ_n^{-1} and τ_p^{-1} be the recombination rates for electrons and holes, respectively. Consider a small volume of the sample through which an electron current I_n is passed. In this volume, charge carriers are generated and may recombine. The excess of carriers depends on the flow of the incoming particles, the number of outgoing particles, and the difference between generated and annihilated carriers.

The rate equations can be easily obtained by considering the flux entering through area A and leaving through area B located at $x = x_0$ and $x_0 + dx$, respectively (see Fig. 3.2). Let J be the flux density of electrons.

The flux of electrons at $x_0 + dx$ is given by

$$J_{nx}(x_0 + dx) = J_n(x) + \frac{\partial J_n(x)}{\partial x} dx \tag{3.4}$$

The net difference in the flux, therefore, is given by

$$[J_{nx}(x_0) - J_n(x_0 + dx)]dy\ dz = \frac{\partial J_n}{\partial x} dx\ dy\ dz \tag{3.5}$$

Thus, the total net increase per unit time in the number of electrons in the volume dx dy dz is $\partial J_n / \partial x$. As the generation and recombination processes are active in this region, the net addition of the electrons is

$$g_n - \underline{n}/\tau_n \tag{3.6}$$

Similarly, the number of holes to be added is

$$g_p - \underline{p}/\tau_p \tag{3.7}$$

This leads to the continuity equation given by

$$\frac{\partial J_n}{\partial t} = -\nabla \cdot J_n + g_n - \underline{n}/\tau_n \qquad (3.8)$$

A similar equation holds for holes, and therefore,

$$\frac{\partial J_p}{\partial t} = -\nabla \cdot J_p + g_p - \underline{p}/\tau_p \qquad (3.9)$$

The solutions of Eqs. (3.8) and (3.9) describe the distribution of electrons and holes as a function of time and space. The photocurrent is measured at two fixed points in space; that is, the space coordinates are fixed. Then the kinetic equations that describe the photoconduction process can be written as [9,10]

$$\frac{d\underline{n}}{dt} = g_n - R_n\underline{n} \qquad (3.10a)$$

and

$$\frac{d\underline{p}}{dt} = g_p - R_p\underline{p} \qquad (3.10b)$$

Equations (3.10) represent the transient behavior of the photocurrent. Variation in the photocurrent originates from the variation in the number of charge carriers (electrons or holes) and not from the variation in their mobility. Therefore, current is replaced by photogenerated charge carriers, $\underline{n}$, in Eq. (3.10). Thus, it becomes a conventional form of the equations often used for examining the kinetics of photoconduction process [9,10].

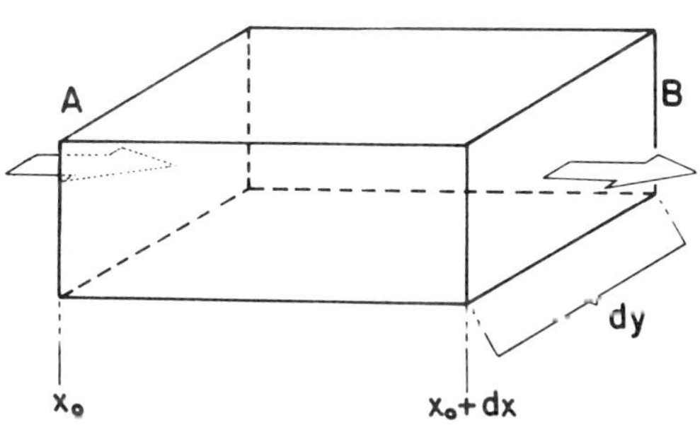

FIGURE 3.2 A small portion of the photoconductor. Electrons enter through surface A and leave through surface B. In addition to this, carriers are created and annihilated in this region.

3.3 TIME CONSTANTS AND RECOMBINATION PROCESSES

The other significant terms in Eqs. (3.9) and (3.10) are the time constants τ_n and τ_p. They are related to the recombination processes and characterize the free charge carrier decay. These processes may be either radiative or Auger type. In general, the recombination process is very complex; many details of its statistics are discussed by Landsberg [22]. Band-to-band recombination is only one of many processes. Recombination takes place through physical defects such as vacancies, dislocations, and grain boundaries. The impurities, which are introduced intentionally to increase the sensitivity, also form the centers for the recombination process. The mobile charge carriers recombine by a variety of mechanisms such as impact ionization, band-to-band recombination, and phonon-assisted recombination. Depending upon the nature of the process, the mathematical expression varies. Here, we will consider simple and commonly observed mechanisms.

According to the recombination statistics, the process is predominantly governed by monomolecular or bimolecular processes or a combination of the two.

3.3.1. Monomolecular Case

In the monomolecular case, the recombination is proportional to the excess of charge carriers, and therefore the rate equation for the rise process is given by

$$\frac{dn}{dt} = g_n - \frac{n}{\tau_n}^{*} \tag{3.11}$$

where g_n is the rate of generation of charge carriers. It is assumed to be constant during the process and is equal to $\alpha I_0 n$, where I_0 is the intensity of radiation generally measured in watts per square centimeter. A similar equation can be written for excess holes. When $g_n = 0$, the equation represents the decay curve. Assuming that τ_n is independent of n, the solution of these equations can be given by

---

*Here we are using the symbols τ_n and τ_p for the relaxation lifetimes related to the recombination process. Unfortunately, the same symbols have been used in the literature to refer to time constants related to the scattering processes.

$$n = \underline{n}_0(1 - e^{-t/\tau_n}) \quad \text{rise} \tag{3.12a}$$

$$\underline{n} = \underline{n}_0 e^{-t/\tau_n} \quad \text{decay} \tag{3.12b}$$

Here, $\underline{n}_0$ is the steady-state value of excess electrons created by optical excitation, and its value can be easily obtained by putting $d\underline{n}/dt = 0$, that is, $g_n \tau_n = \underline{n}_0$. Thus, in the trap-free case and when the diffusion currents are negligible, the rise and fall can be represented by an exponential expression. In this case, the time constant τ_n does represent the average mean lifetime of the electrons (see Chapter 1). This means that it is a free time that electrons spend in the conduction band taking part in the conduction process. Obviously it is a time between two successive collisions with holes and is given by [10]

$$\tau_n = (Cr\ vp)^{-1} \tag{3.13}$$

where v is the relative motion of the electrons with respect to the holes, p is the hole density, and Cr is the "recombination cross section." Ryvkin has defined this as the true microscopic lifetime of the electrons.

The concept and calculation of capture cross section has been developed by Rose [9] for both Coloumb attractive and repulsive potentials. It has been found that the cross section varies from 10^{-12} cm^{-2} for Coloumb attractive to 10^{-20} cm^{-2} for repulsive potentials. It is worth mentioning that the capture cross section is a function of temperature. The large variation in capture cross sections helps to explain the variation in the time constants. I omit further discussion and mathematical details here, as they appear in the literature [see, for example, Refs. 9,23].

In wide-band-gap semiconductors, the density of optically active traps is considerably high, and frequently recombination is dominated by flaws or defects. The concept of carrier capture by a flaw was developed by Blakemore [23]. It has been shown that the continuity equation is also applicable in this situation. The major difference in this case is that the lifetime is governed by recombination through flaws rather than band-to-band transitions.

Assuming that Eq. (3.11) is satisfied and there is no other competing mechanism, the mean lifetime of the electrons is given by

$$\frac{1}{\tau_n} = R_n = (N_t - n_c)\langle c_n \rangle \tag{3.14}$$

where $(N_t - n_c)$ is the number of empty flaws and $\langle c_n \rangle$ is the mean capture cross section given by [23]

$$\langle c_n \rangle = \frac{\int_{E_c}^{\infty} C_1(E)P_e(E)N_e(E)\, dE}{\int_{E_c}^{\infty} N_e(E)P_e(E)\, dE} \tag{3.15}$$

where N_e is the density-of-states function for the conduction band and is defined as [24]

$$N_e(E) = \frac{4\pi(2m_e)^{3/2}}{h^3}(E')^{1/2} \tag{3.16}$$

where m_e is the mass of an electron and E' is the energy measured with respect to the bottom of the conduction band. $C_1(E)$ is the probability of capturing an electron from the conduction band, and P_e is the probability that a state is occupied by an electron.

A considerable variation in two parameters, namely the density of states of recombination centers and their capture cross sections, is observed. The recombination center density varies from 10^{12} cm^{-3} (for high purity and good quality material) to 10^{18} cm^{-3}. Similarly, the capture cross section also varies from 10^{-12} cm^{-2} to 10^{-20} cm^{-2}. Thus, the combination of these two parameters varies over 14 orders of magnitude and so does the photoresponse for a given number of incident photons. Equation (3.15) shows how the mean capture cross section varies with the density of states and with the probability of occupation, which depends on the temperature. Naturally, the lifetime is also a function of temperature.

The above discussion reveals the importance of the density of recombination centers and their capture cross sections for determining the sensitivity of the photoconductors. The details of the mechanisms are already discussed [8–10].

3.3.2. Bimolecular Case

In the bimolecular case, the recombination rate is proportional to the square of the excess of charge carriers instead of directly proportional to it. This is a very familiar Auger process for band-to-band transitions in heavily doped semiconductors. There are extensive calculations of Landsberg [22] on this type of recombination. Even in a pure direct but narrow-band-gap semiconductor, this process was found to be dominant. In this case, the rate equation becomes

$$\frac{d\underline{n}}{dt} = g_n - R_{BM}\underline{n}^2 \qquad \text{(rise)} \tag{3.16a}$$

$$= - R_{BM}\underline{n}^2 \qquad \text{(decay)} \tag{3.16b}$$

Solving these and taking into account the initial conditions, equations for the rise and decay curves are given by [25]

$$\underline{n} = \underline{n}_0 \tanh r(g_n R_{BM})^{1/2} \qquad \text{(rise)} \tag{3.17a}$$

$$\underline{n} = \underline{n}_0/(1 + \underline{n}_0 R_{BM} r) \qquad \text{(decay)} \tag{3.17b}$$

Here again, the value of $\underline{n}_0$ is obtained by putting the rate equation equal to 0, $\underline{n} = (g_n/R_{BM})^{1/2}$. It is clear from these equations that:

1. The rise and decay curves are not symmetrical.
2. The intensity of the equilibrium value of photocurrent is proportional to the square root of the incident radiation.
3. The initial part of the rise curve is directly proportional to time.

None of the above-mentioned properties are considered to be characteristic of the bimolecular process; but the three together can be used to identify the presence of bimolecular recombination processes.

It is also possible that only a fraction, f, of the total electrons are recombined in a linear process and the rest, 1 − f, are recombined in bimolecular processes. Then the rate equation becomes

$$\frac{d\underline{n}}{dt} = g_n - fR_n\underline{n} - R_{BM}(1 - f)\underline{n}^2 \tag{3.18}$$

and the solution is

$$\underline{n} = \underline{n}_0 \frac{[a_1 + (a_1{}^2 + 4b_1 g_n)^{1/2}]}{a_1 + (a_1^2 + 4b_1 g_n)^{1/2} \coth[t/2(a_1{}^2 + 4g_n b_1)^{1/2}]} \qquad \text{rise} \tag{3.18a}$$

$$\underline{n} = n_0 \frac{e^{-a_i t}}{1 + (\underline{n}_0 b_1/a_1)(1 - e^{-a_1 t})} \qquad \text{decay} \tag{3.18b}$$

where

$$\underline{n}_0 = \frac{2g_n}{[a_1^{\;2} + 4b_1 g_n]^{1/2} + a_i}$$

$$a_1 = fR_n, \qquad b_1 = (1 - f)R_{BM}$$

Here we have considered only the predominant and frequently observed recombination processes. In fact, there exist several other mechanisms such as phonon-assisted band-to-band recombination. For example, Landsberg and Robbins [26] have discussed ten types of Auger recombination processes; a detailed study of those for the relaxation of the photocurrent is possible, but for general treatment it is not necessary.

3.3.3. Higher-Order Recombination Processes

In some cases, a power higher than 2 in Eq. (3.16) has also been observed. A typical example is the Auger recombination when the density of photoexcited charge carriers is very high as when the sample is irradiated by a laser beam (photon density is approximately 10^{16}). Junnarkar and Alfano [27] observed a recombination rate given by

$$\frac{d\underline{n}}{dt} = -c\underline{n}^3 - \underline{n} \tag{3.19}$$

It has also been observed [28] that in some cases the recombination rate is proportional to an odd fractional power of $\underline{n}$, for example, 7/3. In such cases, the solution for the specific system must be obtained separately because the general treatment is not available.

Recombination, even band-to-band, is a complicated process, because a variety of different phenomena (e.g., radiative and nonradiative) are involved. The mechanism could include emission or absorption of one or more phonons (in the case of indirect band-gap semiconductors), or recombination could take place via excitons either free or bound to donors (acceptors) as the case may be. Frequently, recombination can also be activated through the traps, dislocations, or defects.

The literature on defects, types of defects, and the parameters that influence the recombination process is extensive [22], and a general treatment is not possible. We will see that even in an apparently simple case the situation becomes very complicated. The theory presented below provides a global picture of the process;

however, if one knows the mechanism of the recombination precisely, then a proper model should be developed for the specific sample.

3.4. EVALUATION OF TIME CONSTANTS: THE SHOCKLEY-READ THEORY

The presence of traps in photoconductors is more normal than their absence, and therefore due consideration should be given to it. In the literature on photoconductivity, the terms "traps" and "recombination centers" are frequently used synonymously. In fact, the difference between them is qualitative. When the probability of recombination is higher than the re-ejection of the charge carriers to the conduction (or valence) band, then the defect centers are called recombination centers. Recombination in the presence of traps involves the statistics of occupation of trap levels, which in turn depend upon the energy of the traps referred to the position of the quasi-Fermi level, which varies with the intensity of radiation. The theory developed by Shockley and Read [29] is reviewed to reveal the role of the trap parameters.

Let us consider a trapping level (independent of its origin) with energy E_t in the forbidden gap. It is neutral when it is empty. It captures one electron from the conduction band and becomes electrically negative. When the hole appears near the occupied trap (or more precisely in the region of influence of the trap), it recombines with the electron. In this process no radiation is emitted; the extra energy of the hole or electron is given to the lattice by the creation of one or many phonons. This is evident as the trapping center is tightly bound to the lattice. Thus the trap promotes the recombination process. This process is generally explained with the help of Fig. 3.3.

The rates of recombination for electrons and holes must be equal and related to the lifetime constant by

$$R_n = R_p = \underline{n}/T_n \tag{3.20}$$

where R_n and R_p are the capture cross sections for electrons and holes, respectively.

The parametric relation between lifetime and other parameters of the material such as the density of the charge carriers in equilibrium and the excess of charge carriers created by incident radiation can be obtained easily by considering the rate at which the charge carriers are captured. Let N_t be the total number of traps. The number of unoccupied traps is $N_t[1 - f(E_t)]$. The rate at which electrons are captured is proportional to $\underline{n}$, and therefore the rate of capture can be written as

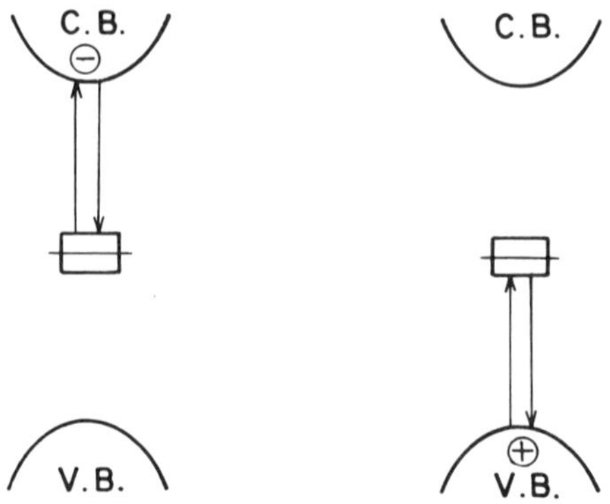

FIGURE 3.3 The fundamental process involved in recombination through traps. (Left) Capture of an electron from the conduction band. (Right) Capture of a hole from the valence band and its return to the valence band.

$$R_{cn} = C_n N_t [1 - f(E_t)] \underline{n} \tag{3.21}$$

where C_n is a constant.

The electrons trapped in the recombination centers do not necessarily meet the hole from the valence band; some of them can be re-ejected to the conductin band. The probability of re-ejection is

$$R_{en} = K_1 N_t f(E_t) \tag{3.22}$$

where K_1 is the recombination constant and its value can be evaluated by considering the equilibrium of the traps,

$$K_1 = c_n \frac{1 - f_0(E_t)}{f_0(E_t)} \underline{n}_0 \tag{3.23}$$

Here subscript 0 stands for equilibrium value. The Fermi distribution function is given by [24]

$$f_0(E) = \frac{1}{1 + \exp(E - E_f)/kT} \tag{3.24}$$

where E_f is the Fermi energy. Therefore,

$$\frac{1 - f_0(E_t)}{f_0(E_t)} = e^{(E_t - E_f)/kT} \tag{3.25}$$

and K_1 becomes

$$K_1 = c_n e^{(E_t - E_f)/kT} \underline{n}_0 \tag{3.26}$$

The net rate at which the electrons are captured is

$$R_n = R_{cn} - R_{en} \tag{3.27}$$

$$= C_n N_t \{\underline{n}_0 [1 - f(E_f)] - n_1 f(E_t)\} \tag{3.28}$$

where n_1 is the number of electrons that would be present in the conduction band if the quasi-Fermi level were taken to coincide with the trapping level and is given by

$$n_1 = 2(2\pi m^* kT/h^2)^{3/2} e^{-(E_c - E_t)/kT} \tag{3.28a}$$

Similarly, the rate of capture of the hole can be estimated. Hole capture rate depends upon the number of holes in the valence band and the number of traps occupied by electrons. Therefore,

$$R_{cp} = C_p N_t f(E_t) \underline{P} \tag{3.29}$$

The rate at which the holes are emitted in the valence band is proportional to the number of traps occupied by the holes and can be written as

$$R_{ep} = k_2 N_t [1 - f(E_t)] \tag{3.30}$$

With similar arguments, the rate of recombination of holes is

$$R_p = R_{cp} - R_{ep} \tag{3.31}$$

$$= C_p N_t [\underline{P} f(E_t) - P_1 [1 - f(E_t)]] \tag{3.32}$$

In the theory proposed by Shockley and Read it is assumed that the excess of electrons $\underline{n}$ is equal to the excess of holes p. In fact, in the presence of traps this assumption is not really justified, as some of the photogenerated electrons are trapped. However, when the density of traps N_t is small compared to the density of photogenerated charge carriers, then $\underline{n} = \underline{p}$ is considered to be a good approximation. In that case the rates of recombination of electrons and holes must be equal. Then

$$R_n = R_p = \underline{n}/\tau_n \tag{3.33}$$

By using Eqs. (3.28) and (3.32), the Fermi function can be written as a function of the capture cross sections C_n and C_p for electrons and holes and the number of charge carriers.

$$f(E_t) = \frac{c_p P_1 + c_n n}{c_n(n + n_1) + c_p(p + p_1)} \tag{3.34}$$

By using Equation (3.32),

$$R_n = R_p = \underline{n}/\tau \tag{3.35}$$

$$= \frac{N_t c_n c_p (np - n_i^2)}{c_n[n_0 + \underline{n} + n_1] + c_p(P_0 + \underline{P} + P_1)} \tag{3.36}$$

since $\underline{n} = n_0 + \underline{n}$ and $\underline{p} = p_0 + \underline{p}$.

Then Eq. (3.36) becomes

$$1/\tau = \frac{N_t(n_0 + p_0 + \underline{p})}{1/c_p[n_0 + n_1 + \underline{P}] + 1/c_n[P_0 + P_1 + p]} \tag{3.37}$$

Dividing both the numerator and denominator by N_t and substituting $\tau_{n0} = [c_p N_t]^{-1}$ and $\tau_{p0} = [c_n N_t]^{-1}$, we get

$$[\tau]^{-1} = \frac{n_0 + p_0 + \underline{p}}{\tau_{p0}[n_0 + n_1 + \underline{p}] + \tau_{n0}[p_0 + p_1 + \underline{p}]} \tag{3.38}$$

where τ_{n0} and τ_{p0} are the lifetimes of electrons and holes in strongly p- or n-type materials, respectively. For small values of $\underline{P}$ the time constant becomes

$$\tau = \frac{\tau_{n0}(p_0 + p_1)}{p_0 + n_0} + \frac{\tau_{p0}(n_0 + \underline{n}_1)}{p_0 + n_0} \tag{3.39}$$

Thus, the general trend of lifetime variation as a function of the Fermi level and as a function of the density of charge carriers can be visualized.

It is very hard to compare Shockley-Read theory with experimental results as it is difficult to find a sample in which all the parameters are evaluated and also there is usually more than one type of trapping level. The second type of trapping level often has a very small hole capture cross section. In such circumstances it is possible, in principle, to extend the theory to a system of two levels, but the obtained mathematical expression will neither be easy to verify nor provide useful additional information about the expected behavior of transient photoconductivity.

The contribution of the second type of traps can be understood intuitively. Electrons captured in the first type of traps would recombined in a normal way. Those captured in the second type of traps take longer to recombine, and meanwhile a large number of electrons from the first kind of traps have recombined. Thus, one observes the two-component rise or decay curve.

There exists an alternative and equally probable approach that is essentially based on the fact that electrons are trapped from the conduction band and then some of them are re-ejected into the conduction band [8]. The consequence of this is that a group of re-ejected electrons take part in the photoconduction process. This delayed contribution affects both the relaxation process and the mean lifetime. The contribution of the delayed component in photoconductivity was recently discussed by Joshi and Martin [12] and will be treated separately.

In the treatment given by Shockley and Read, it is considered that the probability of capture is proportional to the number of empty traps. However, this number is not a constant, even though it is tacitly assumed to be in the theory, but varies with time. The same is true for hole capture and for recombination rates. These arguments eventually alter the expression for the time constant and the form of the relaxation curves.

The Shockley-Read theory and the above-mentioned factors clearly show that the rise and decay curves, in the presence of traps, are very complicated to interpret. Experimental data are rich and include several forms of rise and decay. The literature is extensive, and models to explain those observed forms are abundant. Unfortunately, these models are phenomenological and subjective. The precise role of the local centers (traps and recombination centers are not differentiated at this stage) is not clear, and the parameters of the traps such as density, distribution, and capture cross sections are generally not determined by other independent methods.

Recombination is one of the fundamental processes not only in steady-state and transient photoconductivity, but also in luminescence and photonic devices such as light-emitting and laser diodes. There is a vast amount of literature, both theoretical and experimental, which would occupy a volume in itself. The handling of the recombination information, the complicated and unverifiable mathematical expression for lifetime obtained from Shockley-Read theory, and the limitations involved in it force us to develop a model for the particular experimental data at hand. An excellent discussion on a variety of models is given by Rose [9], Bube [25], and Ryvkin [10]. Developing the basic concepts of recombination through traps, several (or almost all) properties of time-dependent behavior have been satisfactorily explained.

3.5. EFFECTS OF TRAPS ON PHOTOCURRENT

The presence of traps not only makes the photoresponse slow and causes the relaxation curves to deviate from an exponential nature but also causes to vary the photocurrent with some power of intensity of radiation. Rose [9] has considered several models that take into account the effects of a single set of recombination centers and a set of trapping levels, independently, both under low and high radiative excitation. The effects of radiation on the position of the quasi-Fermi level hae also been examined by considering the transition region in which the trapping states act as recombination centers. The general approach is similar to the one adopted by Shockley and Read (i.e., recombination of charge carriers through the centers), but instead of treating the problem in a global form, Rose modeled the process for more specific, experimental conditions. For example, the number of excited electrons and holes is much higher (or lower) than the nmber of electrons trapped in the recombination centers. This and similar views not only helped to visualize the process but also brought to notice the role of several parameters in a particular condition, and further they help to identify experimental situations in which the time constant remains invariant with variation of light intensity and is sensitive to temperature variation. Most important is that the proposed models [9] provide information about the conditions and limitations when lifetime is equal to the response time constant of the photoconductor. Moreover, these models help to visualize the mechanisms in general. Unfortunately, direct verification is not so frequent.

As mentioned earlier, the key parameter in photoconductors is the lifetime of charge carriers, which varies from 10^2 to 10^{-9} sec and determines the responsivity of the photoconductor. The number of ways in which the photocurrent depends on light intensity temperature, and defect states is practically unlimited. The photocurrents have been observed to increase linearly, superlinearly, and sublinearly with light intensity. In some materials, photocurrents have been insensitive to temperature (at least in some range of temperatures), and in others they have been found to increase or even decrease with temperature. Similar behavior has been observed for defect states created by different methods.

A large variety of experimental data increases the tendency to adopt a view similar to one adopted by Rose to explain the observed behavior with the help of a specific model [9] for lifetime constant. Every proposed model has its advantages and limitations but explains the particular observed result. The assumptions involved in the model are not always obvious or verified. A typical example is the model that explains the variation of photocurrent as a noninteger power of radiation. In this case it is assumed that traps are distributed with varying concentration in the upper part of the band gap. As the intensity of the radiation increases, the quasi-Fermi level shifts toward the conduction band edge and an increasing number of traps are converted into recombination centers. Assumptions involved in the process (e.g., trap distribution) are accepted because they are logical, but they are rarely proved by other independent methods. It must be stressed that in a majority of cases there is another possible explanation for the observed results.

A notable example of such models is the explanation of the variation of the photocurrent with non integer power of light intensity. It has been observed that in some cases (for example Sb_2S_3) photocurrent can be expressed as

$$I_{pho} = Cp^{c_2} \tag{3.40}$$

where C is a proportionality constant and c_2 is a constant that lies between 0.5 and 1.

To understand the observed result, it is necessary to assume that the traps are distributed exponentially in the band gap. Even though the model explains the observed phenomenon, it is difficult to verify (and also justify) the approximation. Similarly, with different distributions of traps and with proper parameters, other power dependencies such as $c_2 > 1$ or $c_2 < 1/2$ have been successfully obtained. The approach taken by Rose visualizes a variety of patterns in a simple manner without resorting to elaborate analytic solutions. The time constants obtained are certainly approximate, and the time-dependent behavior of the photocurrent cannot be explained just by fitting these constants at the exponential. This is because the number of electrons or holes, occupied or unoccupied traps, varies continuously with time, and therefore the rate equations should be examined. Such a view has been adopted by Ryvkin [10] and Bube [25].

3.6 EFFECTS OF TRAP PARAMETERS ON THE RELAXATION CURVES

We have seen how the presence of traps greatly alters the kinetics of the charge carriers. The precise way in which the form of the relaxation curves varies with the trap parameters can be understood with the help of the analysis carried out by Ryvkin [10]. He considered several systems in which a few types of recombination centers and traps are present. Because of imprecise differences between trapping and recombination centers, it is very difficult to describe precisely the recombination, trapping, or retrapping processes with adequate assumptions. However, in some special cases and under some approximations, the time constants and also the form of the relaxation curves have been thoroughly examined with a semiquantitative treatment. In order to take into account the variation of charge carriers, traps, and recombination centers with time, their rate equations were developed and solved under certain reasonable assumptions and in separate time domains [10]. Such a solution (a different solution for each time domain) is really not fully satisfactory, but it has been widely accepted because an analytic solution had not been reported so it was the nearest approach so far.

Ryvkin [10] analyzed the effects of traps on the relaxation curves in great detail. It must be realized that even in the case of a relatively simple system, that is, with one type of traps, the description of the processes becomes complicated and the situation can be handled only for special cases. The contribution of the traps or recombination centers depends on the position of the Fermi level and on the density of traps relative to the photoexcited charge carrier density. Ryvkin [10] obtained the mathematical expressions for lifetime as a function of the equilibrium density of charge carriers, the density of photoexcited charge carriers, and the lifetimes of electrons and holes for capture by localized centers for various positions of the quasi-Fermi level (very close to the conduction band, in the upper half of the conduction band, close to the valence band, etc.). The temperature dependence of the relaxation curve and of the time constants is also taken into account through the variation in the probability of the electrons being re-ejected to the conduction band. By considering a large number of extreme cases of recombination through traps (a single type or several types), Ryvkin explained several forms of relaxation curves and, further, obtained semiquantitative relationships between the trap parameters and the lifetime of response time. He also analyzed the effects of traps directly via the form or the relaxation curves. One of the significant models which identify two types of traps can be summarized as follows.

Ryvkin considered a system in which recombination is dominated by recombination centers. Here, the density of recombination centers is so high that the effect of illumination has practically no effect on the empty traps ($N_t - n_c$), where N_t is the density of the total number of recombination centers and n_c is the density of occupied centers. The kinetics of the charge carriers is explained with a set of nonlinear differential equations [10].

The change in the conducting electron density per unit time is given by [10]

$$\frac{d\underline{n}}{dt} = g_n - \alpha_1 \underline{n} - \gamma_n \underline{n} N_t + \gamma_n N_{cm} n_c \tag{3.41}$$

α_1 represents the rate at which the electrons are recombined with the holes in the valence band. Here N_{cm} is the effective density of states in the conduction band reduced to the trap level and is given by

$$N_{cm} = N_c^{-E_t/kT}$$

The sum of the rate of change of electrons in the conduction and in the traps can be written as

$$\frac{d(\underline{n} + n_c)}{dt} = g_n - \alpha_1 \underline{n} \tag{3.42}$$

where $\alpha_1 = 1/\tau_1$.

The simultaneous system of equations given by (3.41) and (3.42) has no analytic solution, and therefore the system is solved in two time domains separately with approximations that are valid only within each separate domain.

Let us consider a rise form of the transient curve, that is, one in which when $t = 0$, both $\underline{n}$ and n_c are equal to zero. At the initial stage, n/τ_1 is small, so Eq. (3.41) becomes

$$\frac{d\underline{n}}{dt} = g_n - \gamma_n \underline{n} N_t + \gamma_n N_{cm} n_c \tag{3.43}$$

Using the initial condition at $t = 0$, $\underline{n}$ and n_c are both equal to zero (here we are considering, for simplicity, a unipolar photoconductor). Therefore,

$$\frac{d(\underline{n} + n_c)}{dt} = g_n \tag{3.44}$$

and hence

$$\underline{n} + n_c = g_n t$$

Substituting this value into Eq. (3.43), its solution can be written as

$$\underline{n} = g_n[\gamma_n n_c[\theta]^2(1 - e^{-t/\theta})] + \gamma_n N_{cm}\theta t] \qquad (3.45)$$

where

$$\theta = [\gamma_n(n_c + N_{cm})]^{-1}$$

This shows that the solution has two components; one is exponential and the other is linear with time. The final form of the relaxation curve depends on the relative values of the parameters. A detailed discussion is given by Ryvkin [10].

For higher values of time, the number of charge carriers increases and the term $\underline{n}/\tau_1$ in Eq. (3.41) is no longer negligible. When $t \gg \theta$, the traps are in thermal equlibrium, and the ratio of the number of excited electrons to the number of traps is given by

$$\frac{\underline{n}}{n_c} = \frac{N_{cm}}{N_t}$$

and hence,

$$n_c = \underline{n}(N_t/N_{cm})$$

(Note that N_t varies with the activation energy of the traps and the temperature.) Substituting this value into Eq. (3.42) gives

$$\frac{d}{dt}\left(\underline{n} + \underline{n}\frac{N_t}{N_{cm}}\right) = g_n - \frac{\underline{n}}{\tau_n} \qquad (3.46)$$

The solution of this equation is

$$\underline{n} = g_n\tau_n(1 - e^{-t/\tau'}) \qquad (3.47)$$

where $(1 - e^{-t/\tau'})$ is given by

$$\tau' = \tau_n(1 + N_t/N_{cm}) \tag{3.47}$$

The expected form of the relaxation curve for this type of trap (known as the α type) is shown in Fig. 3.4 for two time domains.

There exists another extreme possibility. The response time of the electrons may be very small compared to the time required by the centers to capture the electrons. In this case, after shining the radiation, an almost steady state value is obtained, and then a slow process of capturing is noticeable. In the initial relaxation stage, we can neglect the term $\gamma_n N_{cm} N_c$, and therefore Eq. (3.41) becomes

$$\frac{d\underline{n}}{dt} = g_n - \underline{n}\left(\frac{1}{\tau_n} + \gamma_n N_t\right) \tag{3.48}$$

The solution of Eq. (3.48) is

$$\underline{n} = g_n \tau''(1 - e^{-t/\tau''}) \tag{3.49}$$

where

$$\tau'' = (1/\tau_n + \gamma_n N_t)^{-1}$$

Equations (3.47) and (3.49) show the relation between the response times for two different cases.

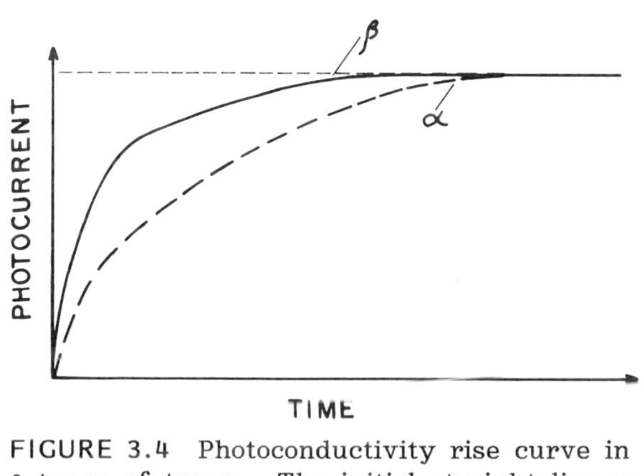

FIGURE 3.4 Photoconductivity rise curve in the presence of α and β types of traps. The initial straight-line portion is slightly exaggerated.

In the second stage of the relaxation process, the kinetics of the traps can be expressed by

$$\frac{dn_c}{dt} = \gamma_n[N_t\underline{n}_0 - n_c N_{cm}] \tag{3.50}$$

The solution of this is

$$n_c = g_n \tau_n (1 - e^{-\gamma_n N_{cm} t}) \frac{N_t}{N_{cm}} \tag{3.51}$$

Substituting Eq. (3.51) into Eq. (3.42) we get

$$\underline{n} = g_n \tau_n (1 - \gamma_n N_t \tau e^{-\gamma_n N_{cm} t}) \tag{3.52}$$

It can be seen that the decay curves are symmetrical with the rise curves. The form of this equation is shown in Fig. 3.4. The mathematical analysis proposed by Ryvkin [10] clearly shows the presence of traps and is reflected by the slow component in the relaxation curve. This treatment has been presented here for the sake of completeness and coherence. Moreover, this is a typical example in which a system of nonlinear differential equations is solved in different time zones. Even though the information obtained is useful for identifying the traps, the solution could be quite different if the initial conditions were altered. A few similar systems have been solved in this manner by Ryvkin [10] to explain several forms of relaxation curves.

With the above-mentioned approach it has been concluded that localized centers can be classified for examining forms of relaxation curves in two different ways depending upon the time required to establish an equilibrium between the occupation of the trapping levels and that of conduction band states. If this time is less than the lifetime, then it is an alpha type; otherwise it is a beta type. In the presence of β traps the relaxation curves show two components. The fast one is very close to the lifetime constant, whereas the slow one depends on the trapping parameters.

The effects of traps are not only on the form of the relaxation curve but also on the steady-state value of the photocurrent in several ways. Here we will briefly delineate a few important characteristics.

1. One intuitively thinks that the presence of traps or recombination centers in the band gap will reduce the sensitivity as a greater number of free carriers will be recombined through

these centers. However, it has been found that the additional number of recombination centers can (but will not necessarily) increase the lifetime of one type of carriers and hence the sensitivity of the photoconductors. Some commercial detectors have been sensitized in this manner. A typical example is the introduction of copper impurity in II—VI compounds such as CdSe.

2. The sensitivity of the device could also be decreased. If the origin of the defect centers is such that it accelerates the recombination process, then the number of free carriers and consequently the photocurrent is reduced. Centers of this type are called "killer centers."
3. The presence of traps can increase the response time considerably, even a few orders of magnitude.
4. Traps and recombination centers influence relaxation curves in several ways depending upon the relative values of their parameters. In practice they can lead to all forms of relaxation curves.

The theories mentioned earlier are certainly valuable when only one type of mechanism is dominant, and in that case the theory roughly predicts only the general form of the relaxation curve.

The major impediment to understanding time-dependent photoconductivity is that the numbers of charge carriers and occupied traps are functions of time during the relaxation process itself. This factor, which enormously complicates the solution of the differential equations, has been overlooked so far [30]. The approximations in the system of nonlinear differential equations, even though they are fully justified, can perturb the solution, and therefore they are made during the final stage rather than the initial stage. Taking these aspects into account, an analytic approach was proposed [30]. This helps to explain some aspects of complicated behavior.

3.6.1. Analytic Approach

Let us consider a simple situation where only one type of defect center (traps) exists in a homogeneous photoconductor. On the basis of extensive theoretical and experimental work, it has been clearly established that impurity and defect states play a predominant role in the recombination (radiative and nonradiative) process.

The mechanisms of the formation of the photoexcited charge carriers and their kinetics can be well understood with the help of Fig. 3.5. When the radiation with an energy greater than the band-gap energy is allowed to impinge on the photoconductor, then

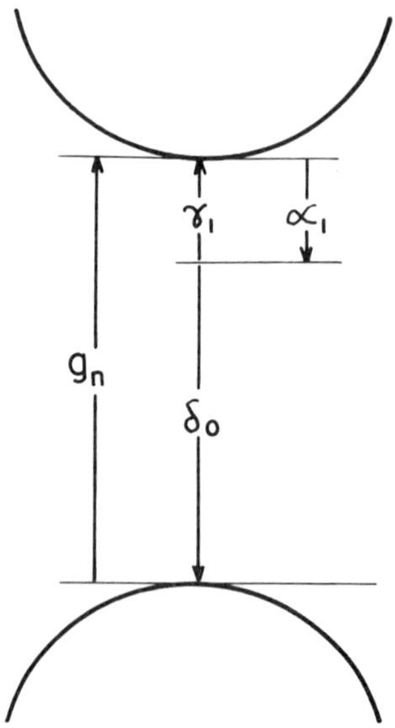

FIGURE 3.5 Basic model showing the transitions involved in the photoconduction process. The system of equations (3.53) is based on this model.

electrons $\underline{n}(t)$ and holes $\underline{p}(t)$ are formed in the conduction and valence bands, respectively. A fraction of the excited electrons are captured by the traps with probability $\alpha_1(N_t - n_c)$. Some of the electrons are re-ejected from the traps to the conduction band with probability γ_1 while others are captured through the nonradiative recombination process with a constant δ_0. This is a typical example where recombination is dominated by defect states or traps [9]. The rate equations under non-steady-state conditions can be written as [11]

$$\frac{d\underline{n}}{dt} = g_n - \alpha_1\underline{n}(N_t - n_c) + \gamma_1 n_c \qquad (3.53a)$$

$$\frac{dn_c}{dt} = \alpha_1\underline{n}(N_t - n_c) - \delta_0 n_c\underline{p} - \gamma_1 n_c \qquad (3.53b)$$

$$\frac{d\underline{p}}{dt} = g_n - \delta_0 n_c\underline{p} \qquad (3.53c)$$

When $g_n = 0$, that is, when the photoexcited charge carriers are not generated, this set of differential equations represents the decay curves.

Equations (3.53a-c) are a typical set of equations involved in the kinetics of the charge carriers in the presence of incident radiation. According to the number of traps and recombination centers, there are distinct ways in which the flow of the charge carriers takes place among them and between the conduction and valence band. Obviously, a different set of equations needs to be constructed for a given situation [e.g., 11]. But in the majority of cases, they would form a nonlinear set of simultaneous equations that are solved either numerically or by approximating some terms in the differential equations [9,11,31]. In doing this, the interpretation of relaxation curves might lose general applicability.

In differential equations, particularly those that are nonlinear and in which the terms are strongly interrelated, small approximations at the initial stages in the equations affect the solution considerably. In many cases, an approximate solution may not represent the observed or true solutions, and it might be far away from the true one. Moreover, the behavior of the solution (numerical or approximate) of a system of differential equations near the equilibrium points is itself a disputable issue and can never represent the time evolution of the photocurrent and it should be examined with the stability theory of Liapunov [30,32] for a set of nonlinear differential equations. This fact has considerable importance and deserves separate consideration.

In order to reduce the errors we solve these equations analytically in a series form and then make approximations at the end [33]. Let us express $\underline{n}$, n_c, and $\underline{p}$ by

$$\underline{n}(t) = a_0 + a_1t + a_2t^2 + \cdots \tag{3.54a}$$

$$n_c(t) = b_0 + b_1t + b_2t^2 + \cdots \tag{3.54b}$$

$$\underline{p}(t) = c_0 + c_1t + c_2t^2 + \cdots \tag{3.54c}$$

The coefficients $a_0, a_1, \ldots, a_n$; $b_0, b_1, \ldots, b_n$; and $c_0, c_1, \ldots, c_n$ can be obtained by the usual methods, and the expressions for them are given below.

$$a_1 = g_n - \alpha_1 a_0 \tag{3.55a}$$

$$2a_2 = -\alpha_1 a_1(N_t - b_0) + \alpha_1 a_0 b_1 + \gamma_1 b_1 \tag{3.55b}$$

$$na_n = -\alpha_1 a_{n-1}(N_t - b_0) + \alpha_1 \sum_i a_i b_{(n-1)-i} \tag{3.55c}$$

The values of a_0, b_0, c_0 are determined by the initial conditions.

Let us consider, for simplicity, the decay curves. That is, a photoconductor is illuminated and when it reaches a steady state the radiation is turned off. Then, using Eqs. (3.54) and knowing the densities of traps, electrons, and holes for t = 0, one can obtain $a_0 = \underline{n}_0$, $b_0 = n_{c_{max}}$, and $c = \underline{P}_{max}$. For these initial conditions the solution n(t) can be written as [33]

$$n(t) = [\underline{n}_0 + \gamma_1 n_{c_{max}} \tau + (\alpha_1 \underline{n}_0 b_1 + \gamma_1 b_1)\tau^2] e^{-t/\tau}$$

$$+ \gamma_1 n_{c_{max}} \tau - (\alpha_1 \underline{n}_0 + b_1 + \gamma_1 b_1)\tau^2 (1 - t/\tau) \qquad (3.56)$$

where

$$\tau = [\alpha_1 (N_t - n_{c_{max}})]^{-1}$$

The other terms are found to be small, and their contributions can be neglected. It is clear from Eq. (3.56) that

1. The presence of traps, as expected, has its predominant effect on the form of the relaxation curves.
2. The response is slow (the response time is greater than the lifetime) because of the presence of the second and the third terms.

It is difficult to visualize the time-dependent behavior from Eq. (3.56) as it depends upon the relative values of trap parameters, photogenerated charge carriers, and the time constant. Therefore, the form is simulated for a given set of values of the parameters and is shown in Fig. 3.6. The parameters used for simulation are given in Tables 3.1 and 3.2. The logarithm of I_{pho} is also plotted versus t (Fig. 3.7) as it is conventionally used to detect the presence of two (or more) components in the relaxation curve.

Figures 3.6 and 3.7 show that the obtained series solution, Eq. (3.56), does explain the previously observed forms of the relaxation curves. Moreover, the slow part of the relaxation curve is a direct consequence of the nature of the system of nonlinear differential equations. With this we want to suggest that the presence of two or more slopes on log I_{pho}Vs t does not necessarily mean the presence of two or more types of traps. A system of single traps through which recombination takes place is enough to show two or more time constants.

TABLE 3.1 Physical Parameters Used for Simulation Purposes [See Eq. (3.56) and Fig. 3.6.]

1. Photoexcited electron—hole pairs (low-level excitation for which relaxation curves are generally nonexponential)	$6.5 \times 10^8\ cm^3/s$
2. Capture cross section	$1.5 \times 10^{-9}\ cm^3/s$
3. Total number of traps	$10^{10}\ cm^{-3}$
4. Total number of occupied traps	$0.33 \times 10^{10}\ cm^{-3}$
5. Lifetime $\tau = [\alpha(N_T - N_{c_{max}})]^{-1}$	0.1 sec

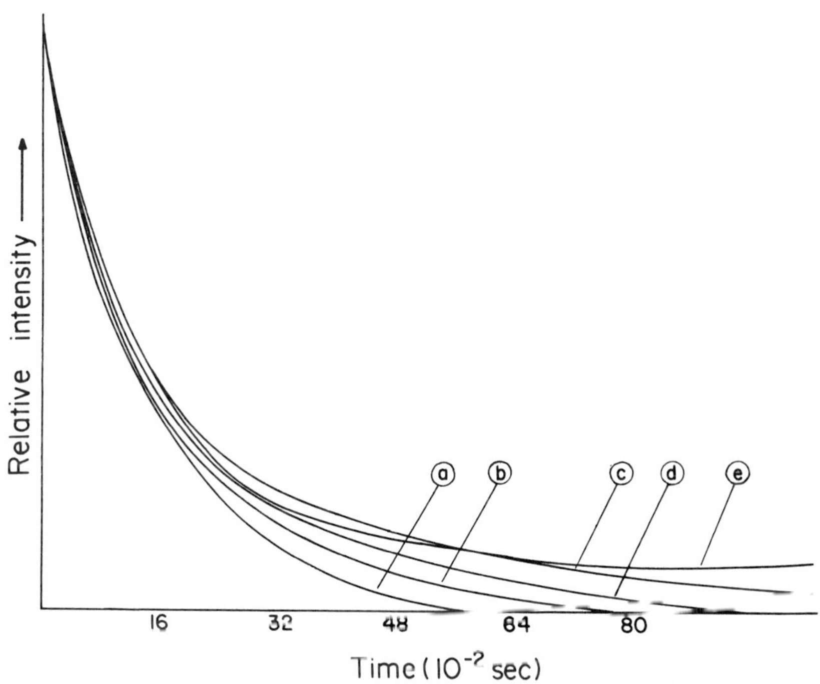

FIGURE 3.6 Relaxation curves simulated by using Eq. (3.56). The values of the parameters used in the simulation are given in Tables 3.1 and 3.2. [After Ref. 33.]

TABLE 3.2 Values of the Variable Parameters Used in Eq. (3.56) and Fig. 3.6

Curve	γ_1	b_1
a	0.4	2.367×10^9
b	0.4	1.183×10^9
c	0.4	9×10^9
d	0.5	7×10^9
e	0.2	7×10^9

The slow component in photoresponse has already been explained on the basis of a phenomenological model [9]. According to Rose, in a decay process, it would take longer to recombine electrons from both the conduction band and the traps than from the conduction band alone. The same is true for the rise curve. More time is needed to inject electrons into the conduction band, because some electrons are trapped in the process. The additional time to fill or empty the traps will be proportional to $N_t/\underline{n}$. Therefore, the observed time is given by [9]

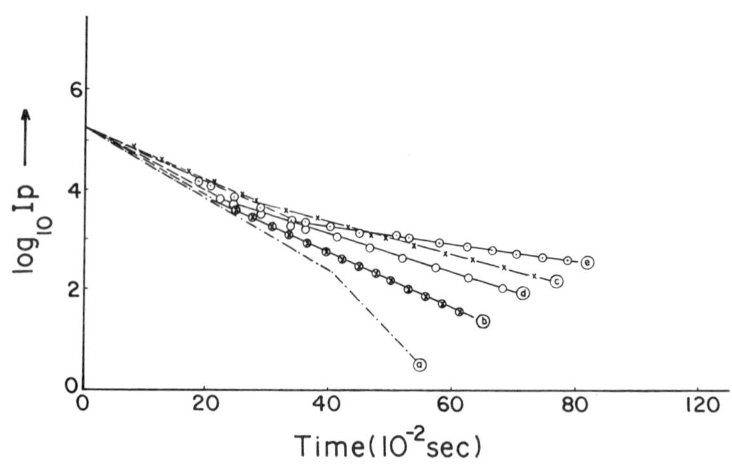

FIGURE 3.7 A plot of the logarithm of photocurrent versus time for the simulated curves. Note that the graph shows two slopes for the curves simulated with one time constant. [After Ref. 33.]

$$\tau_{obs} = \left(1 + \frac{N_t}{\underline{n}}\right)\tau_n \qquad (3.57)$$

However, this is an approximation and is valid only when the traps are not thermally connected to the conduction band and the number of empty traps is large enough that they are not saturated.

The analytic approach presented here [Eq. (3.56)], as expected, shows that the form of the relaxation curves is very sensitive to the relative values of the trap parameters. Meanwhile, the formula obtained from the phenomenological analysis by Rose [Eq. (3.57)] does not take into account the trap parameters but only the ratio $N_t/\underline{n}$. This view, certainly, is not adequate, as the variation in the trap parameters is appreciable. The use of Eq. (3.56) is more proper than the use of Eq. (3.57).

For small values of the time constant (of the order of a millisecond) and the capture cross sections, Eq. (3.56) reduces to

$$\underline{n}(t) = \underline{n}_0 e^{-t/\tau} \qquad (3.58)$$

which is, in fact, a solution of Eq. (3.53a) when the electrons are not re-ejected to the conduction band and corresponds to the trap-free case. A close look at previous experimental results shows that when the lifetime is large, on the order of seconds, the difference between the lifetime and the response time is noticeable; while, if the lifetime is on the order of milliseconds or microseconds, then the difference between them is very small or even negligible. This observation, even though not mentioned explicitly before, can be explained with the present analysis.

Photoresponse in the presence of traps depends on the pre-exponential factor appearing in Eq. (3.56), which in turn depends upon the relative values of the trap parameters, excess of charge carriers, and so on. Thus, the effects of traps on photosensitivity have been taken care of.

The direct effects of traps on the photoconductivity decay curve can very often be understood by solving the system of differential equations in an approximate manner. (See, for example, Ref. 25.) Assuming that the contributions of the first two terms of Eq. (3.53b) are small, the same can be written as

$$\frac{dn_c}{dt} = -\gamma_1 n_c \qquad (3.59)$$

The solution of this equation is

$$n_c = n_{c0} e^{-\gamma_1 t} \tag{3.60}$$

The variation of the density of free carriers during the decay can be approximated by

$$\Delta \underline{n} = -\frac{dn_c}{dt} \tau_{coll} \tag{3.61}$$

where τ_{coll} is the time between two successive collisions. Then, the decay equation for the trap-dominated case becomes

$$\Delta \underline{n} = n_{c0} \gamma_1 \tau_{coll} e^{-\gamma_1 t} \tag{3.62}$$

If there are more than one type of traps, then the resulting equation can be considered the sum of many exponentials and many experimental results could be explained satisfactorily [25].

Equation (3.56) does explain several forms of relaxation curves without considering the presence of two or more types of traps (we do not suggest that the presence of two types of traps is not common, but it should be proved by other independent methods). Thus, the present analysis gives results similar to those of the earlier well-established models [9,10], with the advantage that only one type of localized centers is included in the model and the two components' behavior becomes the consequence of the solution of the system of nonlinear differential equations.

The major success of the present method is that it permits us to develop an alternative view to explain the observed phenomenon, without introducing unverifiable assumptions.

A typical example of this is the explanation of high sensitivity of wide-band-gap photoconductors such as CdSe and CdTe with proper impurity atoms. Their sensitivity along with the response time increases linearly (or almost linearly) with the intensity of the radiation, maintaining the lifetime nearly constant. Many commercially available photodetectors show this property, and a model to explain this behavior has been proposed by Rose [9]. According to the accepted model there is rather uniform distribution of traps in the upper half of the band gap. As the radiation is incident on the photoconductor, the quasi-Fermi level shifts toward the conduction band, converting more traps into recombination centers. This does not alter the lifetime considerably. The number of photoexcited charge carriers is given by the product of their rate of generation and their lifetime. As the latter is unaltered, the sensitivity varies linearly with the intensity of the radiation. The decrease in response time can be understood with the help of Eq. (3.57) as the number of traps is reduced with the excitation intensity. The model proposed by Rose explains the observed behavior. The other details and the numerical estimation were carried out earlier [9].

The same conclusions can be obtained with the help of Eq. (3.56). The lifetime constant is defined for the case in which the recombination is dominated through traps by [23]

$$[\tau_n]^{-1} = (N_t - n_c) \langle c_n \rangle \tag{3.63}$$

It is worth mentioning that the same lifetime constant has been obtained with the present analysis and depends on the reciprocal of the difference between N_t and n_c. Certainly, n_c increases with the intensity of the radiation; however, $(N_t - n_c)$ does not decrease noticeably unless N_t is approximately equal to n_c. Therefore, for the majority of the observed cases, $(N_t - n_c)$ remains practically constant. The above features could be understood with the close analysis of Eq. (3.56) by varying the intensity of the radiation and the parameters of the traps. Thus the use of Eq. (3.56) allows a new possible explanation of the observed experimental results coherently without arbitrary assumptions.

3.7. DETECTION AND SEPARATION OF THE DELAYED COMPONENT

Complications in understanding the form of the relaxation curves are not over yet. Apart from all the factors mentioned here and in the literature, there exists an additional impediment. Electrons from the conduction band are captured by traps according to their capture cross section and remain in the traps for a short time before being re-ejected into the conduction band. Thus, these electrons are delayed with respect to the original ones, and this delay should be reflected in the relaxation process. The delayed electrons evidently prolong the photoresponse time. Since they also have an exponential rise or fall time, the form of the relaxation curve will show the effects as if an extra time constant existed. From the experimental data of the time dependence of photocurrent there is no direct method to detect and analyze the delayed component unless some data processing is carried out.

The inclusion of the delayed electrons leads to a delay differential equation [34], and with the present knowledge solving an isolated equation is in itself a difficult task. Obviously the solution of a system of differential equations involving a delay contribution is out of the question.

Very recently, a delayed component was detected by application of the Fourier transform technique [12]. Let us consider a decay curve of two time constants, say, α_1 and α_2, corresponding to band-to-band and trap-to-valence band recombination. The decay equation can be written as

$$I_{pho}(t) = N_1 e^{-\alpha_1 t} + N_2 e^{-\alpha_2 t} \tag{3.64}$$

where N_1 and N_2 are the constants representing the contributions in the photocurrent decay process from band to band and from traps to band transitions, respectively. Let us consider that both components are in phase and second contribution is not delayed with respect to the first. Then the real and imaginary parts of the Fourier transform are given by

$$\mathrm{Re}(w_i) = \frac{N_1 \alpha_1}{\alpha_1^2 + w^2} + \frac{N_2 \alpha_2}{\alpha_2^2 + w_i^2} \tag{3.65a}$$

$$\mathrm{Im}(w_i) = -\left(\frac{N_1 w}{\alpha_1^2 + w^2} + \frac{N_2 w}{\alpha_2^2 + w_i^2}\right) \tag{3.65b}$$

It is clear that there are no sinusoidal fluctuations in the real and imaginary parts. Now consider a decay curve where the second component is delayed with respect to the first by time τ_0. Then, the decay equation becomes

$$I_{ph}(t) = N_1 e^{-\alpha_1 t} + N_2 e^{-\alpha_2 (t - \tau_0)} \tag{3.66}$$

and the real and the imaginary parts of the Fourier transform of Eq. (3.66) are given by

$$\mathrm{Re}(w_i) = \frac{N_1 \alpha_1}{\alpha_1^2 + w_i^2} + \frac{N_2 \alpha_2 \cos w_i \tau_0}{\alpha_2^2 + w_i^2} - \frac{N_2 w_i \sin w \tau_0}{\alpha_2^2 + w_i^2} \tag{3.67a}$$

$$\mathrm{Im}(w_i) = \frac{N_1 \alpha_1}{\alpha_1^2 + w_i^2} - \frac{N_2 w_i \cos w_i \tau_0}{\alpha_2^2 + w_i^2} - \frac{N_2 \alpha_2 \sin w \tau_0}{\alpha_2^2 + w_i^2} \tag{3.67b}$$

Comparing Eqs. (3.65a) and (3.67a), it can be seen that the real and imaginary parts of the Fourier transform show a periodic variation only when one exponential is delayed with respect to the other.

This is very significant information for detecting a delayed signal in the exponential decay processes, which are very common not only in photoconductivity but also in several other branches of physics (particularly photoluminescence) and in allied fields.

A careful analysis of Eq. (3.67) provides valuable information for analyzing the relaxation curves. The value of α_i and N_1 can be estimated from the first and second terms (w_0, w_1) of the real and imaginary parts. Near t = 0 and t = 1, there is no delay term, and hence the ratio of real to imaginary part is equal to $-\alpha_1/w_1$.

Since w_1 is known (w_n equal $2\pi n$ divided by the number of data points used in the calculation of the Fourier transform), the value of α_1 can be estimated. For t = 0, the real part is N_1/α_1. Thus only two important parameters need to be determined. The numerical estimation of N_2, and τ_0 needs a few trials, but the complete form of the relaxation curve can thus be evaluated quantitatively.

Using this novel technique a complete form of the relaxation curves for high-quality CdTe was successfully analyzed [14]. It is one of the few experiments where the data were recorded in digitized form and the results were analyzed to four-digit (5%) accuracy. Generally, the analysis of the relaxation curves is carried out in qualitative form by plotting a graph of log I_{pho} against time, and the slow and fast components are separated.

This recently developed technique was applied to analyze photoconducting decay curves of a conventional commercial photoconductor, CdSe:Cu [14]. This material is known to have high photoresponse slightly beyond the near-red region and is dominated by impurity states (sensitizing centers) rather than by its intrinsic behavior. Further, it is also known that copper atoms create acceptor states in CdSe [25] and control the mechanism of photoconduction. In this case, it is expected that the transient curve should consist of contributions from the delayed components. Such a study was recently carried out by Joshi and Swanson [14]. The analysis shows that the major part of the photoconductivity curve is delayed by about 140 μs, and it has been suggested that the observed delay is due to copper activation centers.

3.8. TRANSIENT MEASUREMENTS FOR WEAK SIGNAL ANALYSIS

In the previous sections, we discussed the importance of time-dependence measurements of photocurrent, data handling, and processing in the analysis of relaxation curves. Their applications could be also be extended to the detection of weak signals and the separation of two signals of nearly equal amplitude but different frequencies.

3.8.1. Averaging and Detection of Weak Signals

It is well known that the noise associated with optical detectors can be reduced substantially by simply time-averaging the photocurrent. The averaging procedure is carried out by using either a special instrument like a signal averager or an integrate-and-dump circuit. Sometimes the photocurrent is recorded as a function of time (as mentioned in Chapter 2) and then averaged with a suitable procedure. The duration of data collection depends upon the speed of response; for example, for optical communication systems, the averaging procedure is routinely carried out over the bit interval.

3.8.2. Autocorrelation Function for the Analysis of Photosignals

This technique has not been exploited for photoconductivity studies even though it is widely used for radio signal analysis. Its applicability was recently demonstrated by Joshi and Swanson [unpublished work] for separating two signals of nearly the same amplitude but different frequencies. The basic principle is discussed below.

The autocorrelation function, R_{auto}, of a signal X(t), is defined as

$$R_{auto} = \langle x(t) \times (t + \tau) \rangle \qquad (3.68a)$$

$$= \int_{\tau} x(t) \times (t + \tau)\, dt$$

It is clear from this equation that this function gives a measure of similarity between a signal and its delayed version. It is widely used in electrical signal analysis and in communications theory. A typical example is the analysis of a low power signal (frequency and amplitude modulation) for radio communication. The autocorrelation function of a periodic signal is also periodic, as it correlates with itself at a delayed period $\tau, 2\tau, 3\tau$, and so on. Meanwhile, the autocorrelation of a random noise diminishes to zero. Because of this property, the autocorrelation technique is frequently used to detect a signal buried in random noise. This technique has recently also been applied in acoustical measurements, particularly in detection of echoes [35].

A similar situation, namely, a weak signal in a random noise, also appears with added complications in photosignal analysis, where an electrical periodic signal of frequency 58 Hz mixes with a photosignal. The analysis and extraction of such a signal will be a significant achievement in photonic device technology and will have several industrial applications. The issue of weak photosignal detection is becoming more and more crucial and of great consequence, particularly in optical communication.

It is a common experience, independent of the experimental technique employed in photodetection, that the presence of electrical noise of ac frequency is unavoidable in weak signal detection. The noise generally originates from the other laboratory equipment (power supply, stabilizer, etc.), and when its amplitude is roughly equal to the amplitude of the photosignal then analysis and separation of the signal become a tough task. This is particularly observed in cases where a photodetector has a slow response and hence its performance is limited to low chopping frequencies. Phase-sensitive detection cannot be used in these cases. The purpose of the present investigation, therefore, is to employ the autocorrelation technique in this type of situation.

The photosignal at the output of the measuring system can be written as

$$V(t) = V_{pho} \cos wt + V_{electric\ field} \cos(w't + \phi) + V_{noise} \tag{3.69}$$

where w and w' are the angular frequency of the chopper and the electric field (58 Hz), respectively, and V_{noise} is an rms noise. Autocorrelation of V(t) is given by

$$V(t)_{auto} = \frac{V^2_{pho}}{2} \cos wt + \frac{V^2_{electric\ field}}{2} \cos w't \tag{3.70}$$

Equation (3.70) shows that

1. The random noise, V_{noise}, disappears because it is an uncorrelated signal.
2. The periodic components of the signal appear in the correlated function with unaltered frequency.
3. The amplitude of the periodic function in autocorrelation is proportional to the square of the amplitude of the signal rather than its amplitude itself.

The first two features are well known and have already been explored for detecting weak signals in several branches of physics and technology. The third property suggests that this technique can be extended to separte the signals of different frequencies. Even though V_{pho} and $V_{electric\ field}$ are on the same order, their squares are not. It is clear from this equation that autocorrelation should clearly reveal modulation of electric signal on photosignal.

Direct experimental confirmation was obtained recently [Joshi, unpublished work]. A commercially available CdSe:Cu photodetector was used for this purpose. It is known that it has slow photoresponse ($\tau_{response}$ = 1.4 s) because of copper-activated centers.

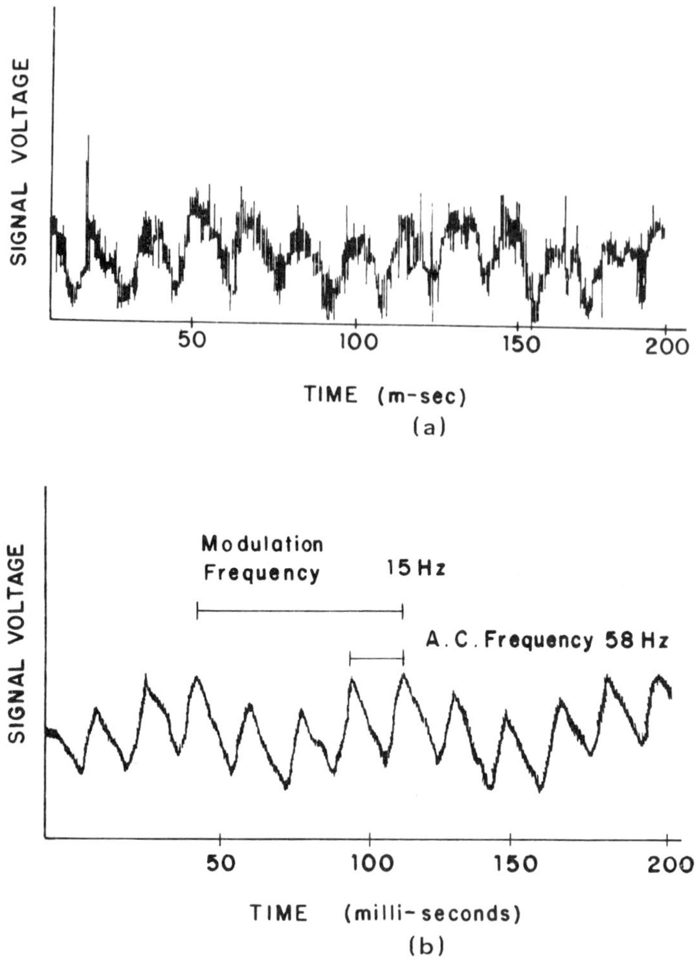

FIGURE 3.8 (a) Photosignal as recorded by the instrument. (b) Autocorrelation of the observed signal. (Unpublished results, permission obtained from Honeywell, Bloomington, Minnesota.)

The detector was illuminated with radiation coming from a white lamp. In order to examine the capability of the present approach, the intensity of the radiation was adjusted so that the photosignal had the same amplitude as the electrical noise. The radiation was chopped with 15-Hz frequency. The photosignal was obtained using a conventional electric circuit. The photosignal was given to the oscilloscope, and was averaged for 10 sweeps before it was then digitized. The sinusoidal form that originated from chopping (15 Hz) is hardly visualized.

Autocorrelation of the photosignal is then carried out with the standard program. The photosignal and its autocorrelation are shown in Fig. 3.8. Figure 3.8b shows two types of oscillations, one corresponding to photosignal (15 Hz) and other corresponding to ac noise. Thus the contributions of two signals of different frequencies are separated.

It is well established that the autocorrelation technique brings out a weak signal from noise. With the help of the above analysis it can be realized that this technique can also separate two periodic signals of different frequencies even though they have nearly the same amplitude. Such situations frequently occur in low-intensity photodetection measurements in the presence of ac circuit noise, a common situation in the laboratory.

Time-dependent photoconductivity measurements could be a powerful technique for understanding the kinetics of charge carriers, the recombination process, and also the response time of a photodetector. They are also valuable tools to detect a weak steady-state signal. However, several types of complications (both theoretical and experimental) mentioned above cast doubts about their general applicability and reliability, at least in the near future, for obtaining information from the relaxation curves. Nevertheless, the recent significant developments encourage us to continue our efforts in the same direction, as a majority of relaxation curves observed in the laboratory are affected by only a few of the parameters discussed above. Contrary to many others, I believe that the efforts to obtain analytical solutions (in series form) and the separation of the delayed component are certainly a rewarding task not only in photoconductivity but also in many time-dependent phenomena.

REFERENCES

1. W. Shockley, Electrons and Holes in Semiconductors. Van Nostrand, Princeton, 1950.
2. I. Tureck, Phys. Stat. Solidi B 62,K45 (1980).
3. P. L. Tyutunnikov, USSR Comput. Math. Math Phys. 20,26 (1981).
4. M. K. Bakhadyrkhanov, T. S. Kamilov, and A. T. Teshabaev, Fiz Tekh. Poluprovodn. 10,328 (1976) [Sov. Phys. Semicond. 10,195 (1976)].
5. R. J. Fleming, J. Appl. Phys. 50,8075 (1979).
6. P. Migliorato, C. T. Elliot, and A. W. Vere, Solid State Commun. 43,307 (1977).
7. A. I. Andrushko, G. G. Kovalevskaya, and S. V. Slobodchikov, Sov. Phys. Semicond. 14,958 (1980).
8. P. McKelvey, Solid State Semiconductor Physics, Harper & Row, New York, 1966.

9. A. Rose, Concepts in Photoconductivity and Allied Problems (Interscience Tracts on Physics and Astronomy, Vol. 19). Wiley-Interscience, New York, 1963.
10. S. M. Ryvkin, Photoelectric Effects in Semiconductors. Consultants Bureau, New York, 1964.
11. S. Wang, Solid State Electronics. McGraw-Hill, New York, 1966.
12. N. V. Joshi and J. M. Martin, Phys. Lett. 113A,318 (1985).
13. A. A. Turnbull, Br. J. Appl. Phys. 15,1051 (1964).
14. N. V. Joshi and S. Swanson, Solid State Electronics, special issue on Electro-Optics, 30,105 (1987).
15. B. H. Schultz, Philips Res. Rep. 16,175 (1961).
16. D. L. Losee, R. P. Khosla, D. K. Ranadive, and F. T. J. Smith, Solid State Commun. 13,819 (1973).
17. R. Chen, S. W. S. McKeever, and S. A. Durani, Phys. Rev. B24,4931 (1981).
18. F. W. Schmidlin, Phys. Rev. B 16,2362 (1977).
19. C. Popescu and H. K. Henisch, Phys. Rev. B 11,1563 (1975).
20. A. Van der Ziel, Solid State Physical Electronics. Prentice-Hall, Englewood Cliffs, N.J., 1971.
21. P. T. Landsberg and A. F. W. Willoughby, Eds., Solid State Electronics, special issue on Recombination in Semiconductors, Vol. 21, 1978.
22. P. T. Landsberg, Guest Editor, Semiconductor Statistics, special issue of Solid State Electronics, 21 (1978).
23. J. S. Blakemore, Semiconductor Statistics. Pergamon, Oxford, 1962.
24. C. Kittel, Introduction to Solid State Physics. Wiley, New York, 1966.
25. R. H. Bube, Photoconductivity in Solids. Wiley-Interscience, New York, 1960.
26. P. T. Landsberg and D. J. Robbins, Solid State Elect. 21: 1289 (1978).
27. M. R. Junnarakar and R. R. Alfano, Phys. Rev. B 34,1045 (1986).
28. A. Haug, Solid State Electronics 21,1281 (1978).
29. W. Shockley and W. T. Read, Phys. Rev. 87,835 (1952).
30. N. V. Joshi, Phys. Rev. B 32,1009 (1985).
31. Y. Marfaing, Photoconductivity and photoelectric effects, in Handbook on Semiconductors, Vol. 2. T. S. Moss, Ed., p. 417, Elsevier Science, Amsterdam, 1983.
32. J. P. La Salle, Stability by Liapunov's Method. Academic, New York, 1961.
33. N. V. Joshi, Phys. Rev. B 27,6272 (1983).

34. R. D. Driver, Ordinary and Delay Differential Equations. Springer-Verlag, New York, 1977.
35. L. E. Turc, M. Claybourn, A. W. Brinkman, and J. Woods, J. Crystal Growth 72,189 (1985).
35. H. Herlufsey, Technical Review (Brüel & Kjaer, NAERUM, Denmark)1:1 (1984).

4

Band Structure Details and Photoconductivity

4.1. GENERAL INFORMATION

Photoconductivity originates from the creation of mobile charge carriers by absorption of radiation of proper energy and their transport up to the electrode. We use the word "proper" and not "energy higher than the band gap" because several semiconductors show photoresponse for energy lower than the band-gap value due to impurities present in the photoconductors. Electronic transitions and the process of creation of mobile charge carriers depend upon the band structure of photoconductors, while their transport depends upon other factors, including the scattering process, mobilities, and recombination mechanisms. In this chapter we will deal only with band structure properties and electronic transitions that generate free charge carriers, which are responsible for the photoconduction process.

Extensive theoretical [1] and experimental [2] work on the band structure of semiconductors has been carried out since 1960 in elemental [3,4], binary [5], and ternary and multinary [6] compounds. Ternary and quaternary alloys are also very frequently used in photodetection technology, and the variation in the band structure as a function of the concentration of one of the constituent elements is known precisely [see for example, Ref. 7]. Band structure properties include not only the magnitude of the band gap and its nature (direct or indirect) but also the valence band splitting, the density of states in the valence band, the positions of donor and acceptor levels in the band gap [8], the formation of donor—acceptor pairs [9] and their nature and binding energies, the presence of quasiparticles such as excitons [10—12] and their complexes [13], and exciton-phonon bound particles [14,15]. All these features related to the band structure are important, as they take part in electronic transitions and, in the majority of cases, in photoconduction also.

The references cited in the previous paragraph are mainly books or review articles rather than research papers. This confirms that the amount of literature on each topic is so vast that it is not possible to cover all the aspects with reasonable detail in one chapter. Here I will bring to notice those aspects that are not very frequently discussed in this context but are potentially important from both theoretical and technological points of view.

It is worth remembering that there are certain cases where the absorption of photons does not lead to the creation of mobile charge carriers. For example, in the far-infrared region the radiation is absorbed by phonons and hence charge carriers are not created.

If the absorption process creates charge carriers, then it is natural to expect a close correspondence between the absorption and photoconductivity spectra. However, there are certain circumstances where one-to-one correspondence is not observed and therefore only the gross features of the spectra are compared. Some examples follow.

1. Mobile charge carriers, created in the absorption process, are recombined before they reach the electrode. A typical example is a surface recombination that reduces the photoresponse on the higher-energy side of the main peak.
2. Transition metal impurities are introduced into a photoconductor. It is well known that 3d electrons, even in solids, maintain their localized character [16—18] and the absorption originating from the transitions between these states will not be reflected in the photoconductivity unless the final state is very close to or thermally connected to the conduction band.
3. The absorption takes place through the transitions between two defect states located within the band gap. In this case, electron—hole pairs are created, but they are not free and hence will not contribute in the photoconduction process.

Excluding the above-mentioned situations, the photoconductivity spectral response can be best understood if we know the photon absorption processes in semiconductors. Therefore, a brief survey of these processes is presented here.

4.2. SURVEY OF ABSORPTION PROCESSES

4.2.1. Fundamental Absorption Edge Region

In the edge region, the absorption of photons takes place by a transition in which an electron from the valence band is excited to the conduction band and the absorption coefficient is very high in this region, on the order of 10^4 to 10^5 cm^{-1}.

There are two types of transitions, direct and indirect; in the latter the absorption is associated with the creation and annihilation of phonons. The presence of phonons in these cases is needed to conserve the momentum of the system, and the details of this process will be discussed later. In direct transition, initial and final states have the same crystal momentum (i.e., $\underline{k} = 0$) and can be visualized with the help of Fig. 4.1. Here it is worth mentioning that in the present discussion we are ignoring the band tailing that perturbs the band structure near $\underline{k} = 0$. The effects of doping are also neglected.

The difference between direct and phonon-assisted transitions originates from the band structure of the semiconductor. For those semiconductors, the valence band maximum and the conduction band minimum have the same wave vector, the momentum is balanced, and the presence of phonons is not needed. Typical examples are CdSe, CdTe, InSb, and PbS. In other cases, the valence band maximum and the conduction band minimum have different $\underline{k}$ values, and hence the momentum is not conserved unless phonons are generated or annihilated. Examples of the second type are Ge, Si, and GaP. The presence of phonons is impressively demonstrated by the optical

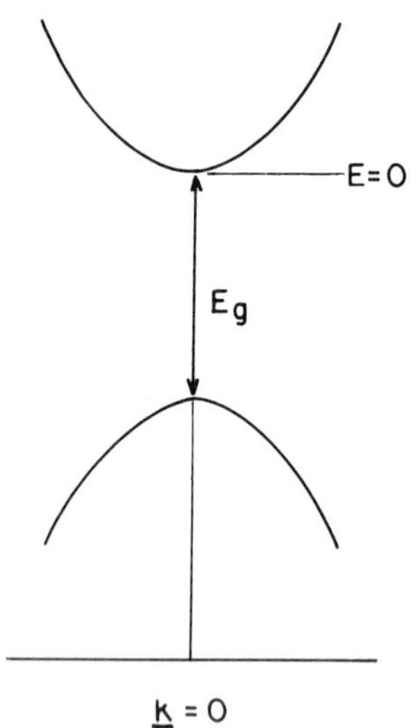

FIGURE 4.1 E–$\underline{k}$ diagram showing a direct optical transition at $\underline{k} = 0$. The impurity states near the band edges are not considered here.

absorption edge in a single crystal of GaP. At low temperature (77 K), the absorption edge clearly shows the contribution from the various phonon modes and their combinations [longitudinal optical (LO), longitudinal acoustical (LA), longitudinal transverse (LT), optical acoustical (OA), and optical transverse (OT)] [19]. This is a process in which additional particles are involved. Obviously, the transition probability has to be different, and hence the absorption coefficient and the shape of the fundamental absorption edge are different. The mathematical formalism and its output for direct and indirect band gaps are altered. The details are available in several books (see, for example, Refs. 2–5).

We consider here a very simple approach that explains the pertinent aspects of the photoconductivity process. In an absorption mechanism, both energy and momentum of the system are conserved. Conservation of crystal momentum is also a consequence of quantum mechanics [3,4]. Let us first consider the implications of momentum conservation.

Let $\underline{P}$ be the momentum of the incident photon, with its magnitude given by h/λ, where h is Planck's constant and λ is the wavelength of the incident photon. According to the conservation of momentum,

$$\underline{P}_2 - \underline{P}_1 = \bar{i}\left(\frac{h}{\lambda}\right) \tag{4.1}$$

where $\bar{i}$ is the unit vector along the direction of the photon. $\underline{P}_1$ and $\underline{P}_2$ are the crystal momenta associated with the initial and final states of the electron $\underline{P}_1$ and are given by the expression

$$|\underline{P}_1| = h/\lambda_1 \tag{4.2}$$

where λ_1 is the de Broglie wavelength of the electron, which is easy to estimate (at 300 K, λ, is approximately 5×10^{-7} cm). The wavelength of the radiation in which we are interested is on the order of 10^{-4} to 10^{-5} cm (visible and infrared regions), and for that value, the right-hand side of Eq. (4.1) is very small compared to the left-hand side. Therefore, we can assume $\underline{P}_2 = \underline{P}_1$. Since the momenta of the initial and final energy states are the same, there is no need for the third particle (phonon) to conserve it.

The form of the absorption edge and the structure near the band gap can be understood by examining the variation in the absorption coefficient as a function of wavelength (or energy). Photon absorption depends on the transition probability under a perturbing electromagnetic field of incident radiation, which is generally expressed with operator $\hat{O}$. The matrix element of the transition probability between initial state i and final state f can be written as

$$M = \int \psi_i \hat{O} \psi_f \tag{4.3}$$

where ψ_i and ψ_f are the initial and final wave functions, respectively. The operator $\hat{O}$ and the wave functions can be written as

$$\hat{O} = A \exp[2\pi i(\bar{r} \cdot \bar{i})/\lambda] \tag{4.4a}$$

$$\psi_i = U_{k_1}(r) \exp[i\bar{\underline{k}}_1 \bar{r}] \tag{4.4b}$$

$$\psi_f = U_{k_2}(r) \exp[i\bar{\underline{k}}_2 \bar{r}] \tag{4.4c}$$

where $U_{\underline{k}_1}(r)$ and $U_{\underline{k}_2}(r)$ are the crystal potentials. The matrix element contains the term exp $\{o[\underline{k}_1 - \underline{k}_2 + (2\pi/\lambda)\bar{i}]\cdot\bar{r}\}$. This is a periodic function and varies rapidly unless $[\underline{k}_1 - \underline{k}_2 + (2\pi/\lambda)\bar{i}] = 0$. Multiplying by h, we get the equation that represents the conservation of momentum. In short, in this case $\underline{k} = 0$; that is, only vertical transitions are allowed (see Fig. 4.1). For $\underline{k} = 0$, a photon with a minimum energy will be absorbed according to the equation

$$h\nu = E_c(\underline{k}) - E_v(\underline{k}) \tag{4.5a}$$

$$h\nu = \Delta E = E_g \tag{4.5b}$$

where E_c and E_v represent the energy of the electrons in the conduction and valence bands, respectively. Thus, an absorption edge begins at the wavelength corresponding to energy E_g, provided that no other types of transitions are present in this region. The impurity states often make their appearance near the band edge, and this should be taken into account in interpreting the results.

The possibility of the absorption of photons in the frequency interval $d\nu$ depends upon the transition probabilities between the initial and final states and the density of the number of states available in the valence band with energy lying between $-(E_g + E')$ and $-(E_g + E' + dE')$.

For the flat conduction band,

$$E_c(\underline{k}) = \text{constant} \tag{4.6}$$

and

$$h\, d\nu = dE' \tag{4.7}$$

The number of transitions, N_{trans}, per unit of time, in the interval ν and $\nu + d\nu$ is given by

$$N_{trans}\, d\nu = P_k N(E')\, dE' \quad (4.8)$$

where P_k is the transition probability for a particular $\underline{k}$. The density of states in the small energy interval E' and E' + dE' is gien by

$$N(E') = \text{constant} \times (E')^{1/2} \quad (4.9)$$

α, the absorption coefficient as a function of energy becomes

$$\alpha = \text{constant} \times P_k \times (h\nu - E_g)^{1/2} \quad (4.10)$$

The above formula is obtained under the assumption that the valence band is derived from the s (l = 0) wave function of the individual atoms and the conduction band from p (l = 1) wave functions. However, this is not always strictly true. Recent investigations on copper-based ternary semiconducting compounds [6] and diluted semimagnetic semiconductors show that in these compounds the valence band contains a mixture of 3d levels of copper and 3d levels of the transition metal ions, respectively. In this case, for $\underline{k} = 0$, the transition probability is not high, but as $\underline{k}$ departs from 0 the transition probability is proportional to $\underline{k}^2$, that is, $h\nu - E_g$.

In parabolic band approximation the absorption coefficient can be given by

$$\alpha = C_1(h\nu - E_g)^{3/2} \quad (4.11)$$

where C_1 is a constant whose value can be obtained by using a quantum-mechanical treatment (see, e.g., Ref. 8) and is

$$C_1 = \frac{2}{3}\left[\frac{\pi e^2}{\varepsilon_i C h^3 m\, \varepsilon_0}(2m_r)^{3/2}\right]\left[2\frac{m_r}{m}\right]\frac{1}{h\nu} f_{ij}$$

where f_{ij} is the oscillator strength of the transition, m_r is the reduced mass of the electron–hole system, and ε_i is the real part of the refractive index.

Here it is assumed that the density of the defect states near the bottom of the conduction band and near the top of the valence band is very small. In amorphous semiconductors the density of states is different from that mentioned above and the momentum is not conserved. For these reasons, this formalism is not valid for amorphous materials. Naturally, the absorption edge and the structure associated with it will show different behavior (sometimes radically different) which will also be reflected in the photoconductivity spectra. This will certainly have significant consequences in photodetection technology.

4.2.2. Absorption Due to Indirect Transitions

When the initial and final states of the electron possess different values of crystal momentum, that is, when P_i and P_f have different values, then the momentum is not conserved. In order to balance it, an additional particle or quasiparticle must be emitted or absorbed. It is obvious that in the present system, momentum and energy will be transferred to the crystal lattice, that is, to phonons—quasi-particles of lattice vibrations. In this case, the photon will be absorbed with a minimum energy given by

$$h\nu = E_g - E_{ph} \tag{4.12}$$

where E_{ph} is the energy of the phonon. That is, the photon takes energy from the phonon to make an indirect band-gap transition. Figure 4.2 shows the details of the process in which the phonon is either emitted or absorbed. The precise calculation of the absorption coefficient in phonon-assisted mechanisms is rather complicated; its numerical estimation has been carried out by Bardeen et al. [8].

Phonon-assisted transitions are explained through the presence of a virtual state with a very short lifetime. First an electron is excited from the top (or very near the top) of the valence band to the bottom of the conduction band (same value of $\underline{k}$; see Fig. 4.2, transition 1). The electron makes the second transition to the lowest possible energy level of the conduction band with the emission or absorption of a phonon of momentum almost equal to the difference between two electronic states.

In indirect transitions electrons and holes are created in the conduction and valence bands, respectively, and hence these transitions are also observed in photoconductivity. Several high-speed and high-responsivity photodetectors are available from indirect band gap semiconductors (e.g., Si, Ge).

The precise calculation of the magnitude of the absorption coefficient is rather complex because of the contribution of different types of phonons (longitudinal, transverse, acoustical, optical, etc.). However, the shape of the absorption curve can be visualized in a relatively simple way.

Let α_i be the indirect transition absorption coefficient, which can be written as [3]

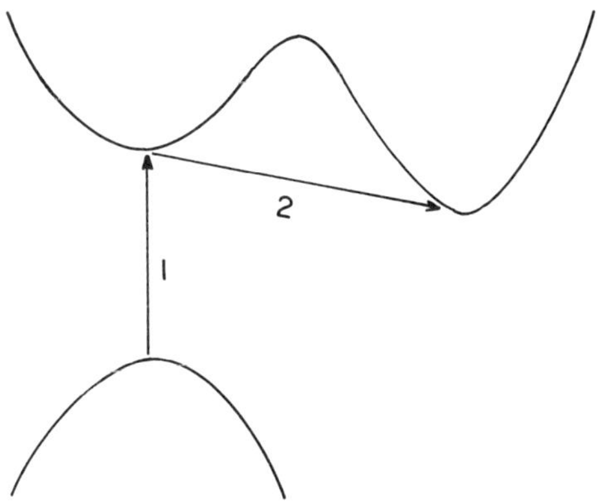

FIGURE 4.2 E–k diagram showing indirect transitions.

$$\alpha_i = \alpha_a + \alpha_e \tag{4.13}$$

where α_e and α_a are the contributions from phonon emission and absorption processes, respectively. As the energy is conserved in both processes, the absorbed photon can be written as

$$h\nu = E_g \pm E_{ph} + E + E' \tag{4.14}$$

where E_{ph} is the energy of the phonon, E is the energy of the electron in the conduction band, and $-(E_g + E')$ is the energy of the electron in the valence band. Note that the latter is measured with respect to the bottom of the conduction band. The number of states in an energy range between E and E + dE is given by

$$N_c(E)dE = C_2(h\nu - E_g \mp E_{ph} - E')^{1/2}\, dE \tag{4.15}$$

To calculate the number of pairs of states that give the energy between E and E + dE, it is necessary to integrate a portion of the valence band. The upper limit is decided by the energy of the phonon and is given with the help of

$$E'_m = (h\nu - E_g - E_{ph}) \qquad \text{phonon emission} \tag{4.16a}$$

$$E'_m = (h\nu - E_g + E_{ph}) \qquad \text{phonon absorption} \tag{4.16b}$$

The density of the states in the valence band, which gives the transition in the range dE or in the frequency range ν to $\nu + d\nu$ is proportional to $N_\nu(E)\,dE$. The equation representing the number of states becomes

$$N(\nu)\,d\nu = C_3 h\nu \int_0^{E_m} (E'_m - E')^{1/2} E'^{1/2}\,dE' \tag{4.17}$$

The absorption coefficient is proportional to E'^2_m.

The emission and absorption of phonons depend on the number of phonons present at a particular temperature, and this number is given by the expression

$$\bar{N}_{pho} = [\exp(E_{ph}/kT) - 1]^{-1} \tag{4.18}$$

The contribution to the absorption, therefore, is given by

$$\alpha_a = c_4 \frac{[h\nu - E_g + E_{ph}]^2}{\exp(E_{ph}/kT) - 1} \tag{4.19a}$$

Similarly, the contribution from the emission of phonons process is

$$\alpha_e = c_5 \frac{[h\nu - E_g - E_{ph}]^2}{1 - \exp(E_{ph}/kT)} \tag{4.19b}$$

for $h\nu > E_g - E_{ph}$

(Note: Phonons are bosons and therefore follow Bose–Einstein statistics. The ratio of the probabilities for phonon emission and absorption is given by N_{p+1}/N_p.)

The absorption coefficient for indirect transition is given by

$$\alpha_{total} = \frac{c_6}{E_g^2}\left\{\frac{(h\nu - E_g + E_{ph})^2}{\exp(E_{ph}/kT) - 1} + \frac{(h\nu - E_g - E_{ph})^2}{1 - \exp(E_{ph}/kT)}\right\} \tag{4.20}$$

for $h\nu > E_g - E_{ph}$

The detailed calculations are given elsewhere (see for example Refs. 3 and 4 and references therein). Here only brief and necessary information is provided at the expense of a precise mechanism. The constants involved in these equations can be evaluated rigorously on quantum-mechanical grounds. In the present simplified treatment I have intentionally omitted the discussion of these constants. Importance is given to the energy dependence of the absorption coefficient near the band-gap region in both direct and indirect semiconductors. This aspect is crucial for photogeneration of charge carriers. In fact, in the above discussion it is necessary to consider the contribution from the longitudinal and transverse optical and acoustical modes. However, such treatment is beyond the scope of this chapter.

Electronic transitions will take place not only at $\underline{k} = 0$ but also at $\underline{k}$ different from 0. The variation of the absorption coefficient near the band gap can be summarized as follows.

For direct allowed transitions,

$$\alpha = A_1(h\nu - E_g)^{1/2} \tag{4.21a}$$

where A_1 is given by

$$A_1 = [\pi e^2(2m_r)^{3/2}/nch^3m\,\varepsilon_0]$$

For forbidden direct transitions,

$$\alpha = A_2(h\nu - E_g)^{3/2} \tag{4.21b}$$

where A_2 is given by $A_2 = (2/5)A_1(2m_r/m)1/h\nu$.

For indirect transitions between indirect valleys,

$$\alpha = \frac{A_3}{E_g^2}\left\{\frac{(h\nu - E_g + E_{ph})^2}{\exp(E_{ph}/kT) - 1} + \frac{(h\nu - E_g - E_{ph})^2}{1 - \exp(E_{ph}/kT)}\right\} \tag{4.21c}$$

The value of the constant A_3 varies with the type of phonon that contributes in the absorption process.

4.2.3. Absorption and Photoconductivity Due to Transitions from Localized States Within the Band Gap

For energy less than the band-gap energy E_g, absorption as well as photoresponse are observed that are attributed to the transitions originating from the localized energy states (acceptors and donors) within the band gap. The photosensitivity of semiconductors is increased by introducing a proper impurity, and in such cases the response starts slightly below the band gap. Plenty of literature is available [2—5] on this aspect, and therefore we avoid unnecessary repetition.

a. Donor-Acceptor Pairs

The transitions from donors or acceptors are very common; it can also happen that in compensated materials electrons-holes form pairs. They act as stationary molecules imbedded in the host crystal. Since the separation between them is not very great, there exists a measurable Coulomb interaction, and the combined energy level is shifted. The magnitude of the shift is given by Coulomb interaction and is expressed by [20]

$$E_{\text{ele-hole}} = \frac{e^2}{\varepsilon R} + J \tag{4.22a}$$

where J is known as the Coulomb integral and represents the cohesive energy of the excited electron—hole pair. Thus, the energy of the pair in a semiconductor of band gap E_g is given by

$$h\nu = E_g - |E_D + E_A| + \frac{e^2}{\varepsilon R} + J \tag{4.22b}$$

The energy of the pair, varies with the distance R, between donor and acceptor. If the phonons are created or absorbed in the donor—acceptor formation process, then phonon energies should be included in Eq. (4.22b).

Sometimes the unique identification of electron-hole pair is rather difficult because the value of R is not specific and in the same energy region there are other optical transitions that are equally possible.

Equations (4.22) are obtained with the first-order perturbation theory; however, appreciable changes have not been reported with the second-order perturbation. These transitions are frequently observed in compensated materials, and hence absorption and photoluminescence spectroscopy [20] provide information about the distance between donor and acceptor. In compound semiconductors it is very useful to know at which site (cation or anion) a particular impurity atom is located.

These transitions are associated with the localized complex and cannot create free charge carriers. Hence, electron—hole pair transitions will not appear in the photoconductivity spectrum. Therefore, those transitions that are absent in photoconductivity and present in absorption (or photoluminescence) most probably have their origin in donor—acceptor pairs.

4.2.4. Exciton and "Exciton Complex" Photoconductivity

b. Exciton Photoconductivity

In an absorption process, electrons and holes are created, and they move freely in the crystal. It can also happen that under a Coulombic attractive force, electrons and holes form a two-particle system (similar to a hydrogen atom, an electron revolving around the nucleus) with quantized energy levels and move together as a unit through the crystal lattice. This system of quasiparticles is known as an exciton [21].

Formal exciton theory was discussed extensively in 1960 [22,23], and the presence of excitons has been confirmed both theoretically and experimentally, by several workers in different materials [24, 25]. The energy states of the quasiparticle have been rigorously obtained by using quantum mechanics [24]; however, the same result can be demonstrated with the extension of the hydrogen atom (two-particle system).

The energy states of an exciton system can be given by [26]

$$E = -\frac{R_\infty}{n^2}\left(\frac{\varepsilon_0}{\varepsilon}\right)\left(\frac{m_r}{m_e^*}\right) \tag{4.23}$$

where R_∞ is the Rydberg constant and m_r is the reduced mass given by

$$[m_r]^{-1} = [m_e^*]^{-1} + [m_h^*]^{-1} \tag{4.24}$$

m_e^* and m_h^* are the effective masses of electrons and holes, respectively, which for simplicity are considered scalars instead of second-order tensors.

It is clear that excitons have an infinite number of energy states leading to the continuum where electrons and holes are free, that is, in the conduction and valence bands, respectively.

Thus, zero energy denotes the bottom of the conduction band, and the energy of the exciton should be measured with respect to it. Exciton spectra have been reported in several semiconductors and insulators [10,11]. The energy diagram of the exciton is shown in Fig. 4.3.

The separation between the energy states and their positions with respect to the conduction band is similar to the separation between energy states originating from impurity atoms. In order to emphasize the fact that exciton states are not localized, the X axis of Fig. 4.3 represents the position of the exciton in the crystal. In fact, the exciton has its wave vector $\underline{k}_{exciton}$, and the energy corresponding to it is given by [26]

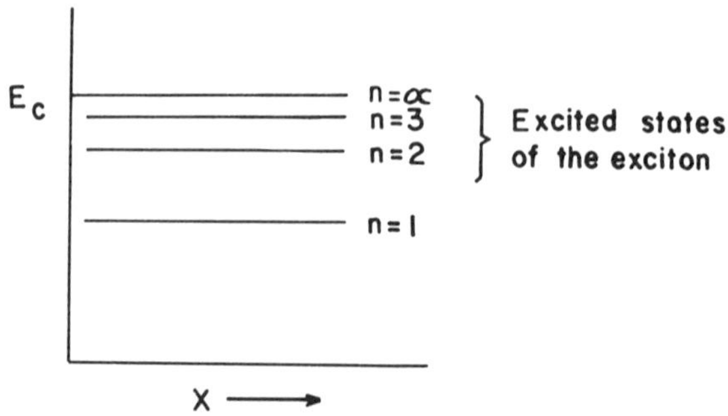

FIGURE 4.3 Energy scheme for the exciton in the crystal.

$$E = -\frac{\hbar^2 k^2_{exciton}}{2(m_e^* + m_h^*)} \tag{4.25}$$

In cases where excitons are created without phonon generation, then $\underline{k}_{exciton}$ is very small and the energy corresponding to it can be safely neglected.

Experimental studies of optical absorption spectra recorded at low temperature (even at 77 K) show a remarkably good agreement between theory and experimental data. The exciton spectrum with several spectral lines (n = 1,2,3,4) has been observed [25]. In some materials, such as Cu_2O, the transition corresponding to n = 5 has also been detected (Fig. 4.4) [27]. From spectral studies the value of the Rydberg constant for excitons has been reported. It is clear from Eq. (4.25) that exciton structures lie very close to and on the low-energy side of the main absorption edge.

As mentioned earlier, the exciton, a quasiparticle composed of an electron—hole pair, is electrically neutral and moves in a crystal as a single unit. Therefore the movement of an exciton should not carry charge and its presence could not be reflected in the photoconductivity spectrum. However, several experimental studies reveal that in some semiconductors the exciton structure does appear in the photoconductivity spectrum [28,29]. A typical example is shown in Fig. 4.5a, where A, B, and C are the structures associated with the valence band structure of CdS as shown in Fig. 4.5b. This apparent anomaly can be understood with the following considerations.

Free excitons can be ionized by interaction with acoustic phonons or with impurities (neutral or ionized) [30]. One of the probable mechanisms is the Auger process by means of which an exciton dissociates into free electrons and holes [31]. An exciton can also dissociate in an electric field produced by an ionized impurity or space

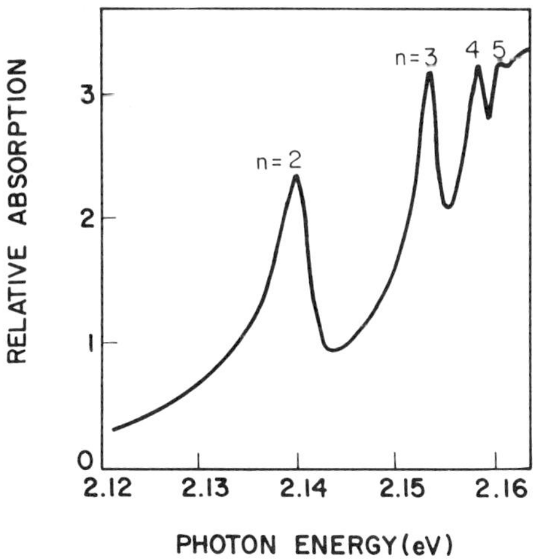

FIGURE 4.4 Exciton absorption spectrum recorded at 77 K in Cu_2O. [After Ref. 27.]

charge created at the surface [32]. Depending upon the material, the exciton may dissociate with any of the above-mentioned processes. The details of this procedure in a given system are not clear, but what is known is that an exciton can be annihilated, giving rise to a free electron and a free hole in the conduction and valence bands, respectively. Under these circumstances, the photoconductivity spectrum reveals the exciton structure, and in several semiconductors, particularly in II-VI compounds such as CdS and CdSe, this structure has been reported [25,33].

It is evident that the exciton appears as an absorption maximum on the low-energy side of the absorption edge. Similarly, in the photoconductivity spectrum one expects a maximum on the low-energy side of the principal absorption edge, but this is not always the case. There are two types of materials. In the first type, the photoconductivity maximum corresponds to the maximum in the absorption [34], whereas in the second type a photoquenching or dip is observed [35]. This peculiar behavior is explained [36] by considering the fact that the values of the transport parameters such as mobility and lifetime are noticeably reduced at the surface, where the exciton absorption takes place.

Excitons are often created with one or more longitudinal optical phonons. Theoretical calculations are supported by many experimental observations. When several phonons are created, the exciton is called a hot exciton; the mechanisms of formation and annihilation are discussed by Planel et al. [37]. In photoconductivity, the cre-

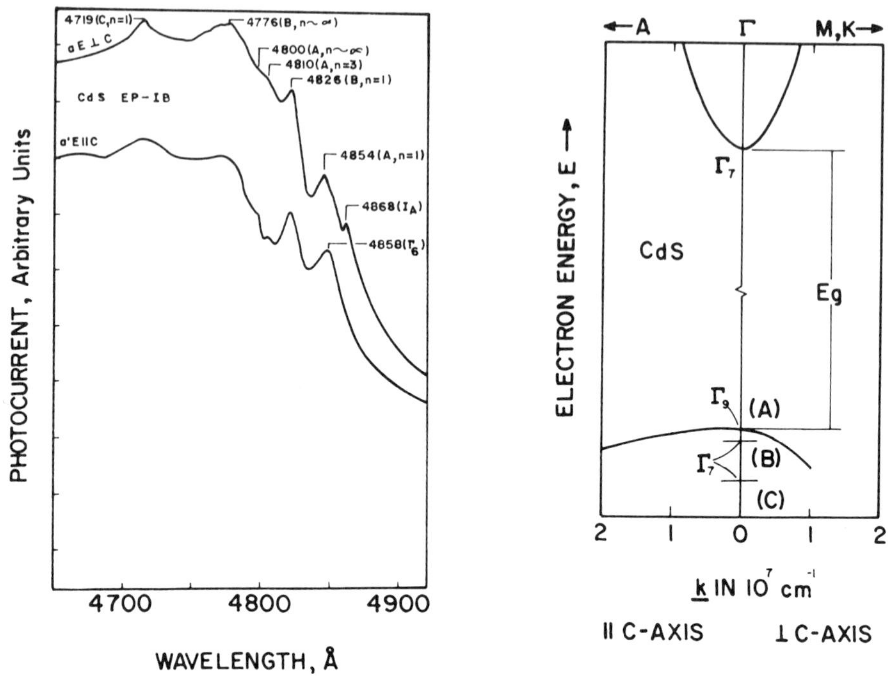

FIGURE 4.5 Exciton structure in CdS recorded at 77 K. [After Ref. 33.]

ation of phonons is reflected by several dips separated by the phonon energy, $\hbar\omega$. The positions of the peaks can be given by [38]

$$h\nu = E_{exciton} + n\hbar\omega_0 \tag{4.26}$$

where $E_{exciton}$ is the energy corresponding to the free exciton. Thus, it produces oscillations in the photoconductivity spectrum, the number of oscillations depending on the number of phonons created in the process. A typical example is shown in Fig. 4.6.

The oscillatory behavior is also attributed to the variation of the transport properties such as lifetime and mobilities of the charge carriers, when excitons are annihilated. A convincing proof of the latter is obtained by phase and photo-Hall measurements [38,39] (see Fig. 4.7). The experimental results do indeed confirm the variation of mobilities and lifetime with the minima in the photoconductivity spectrum.

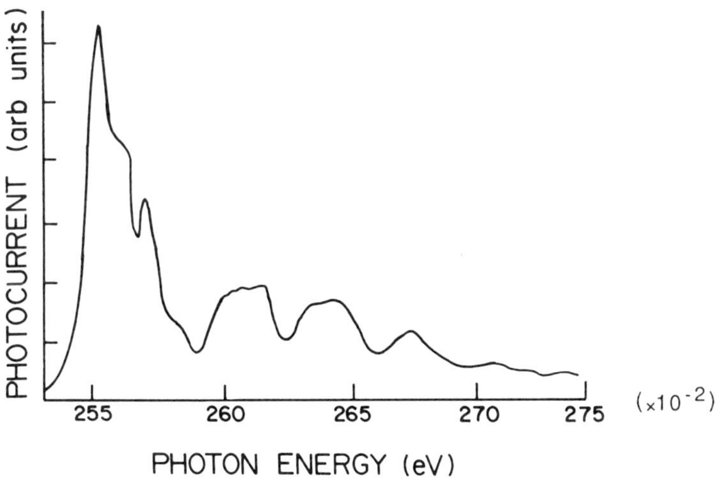

FIGURE 4.6 Oscillatory photoconductivity reported in CdS. Electric field is perpendicular to the C axis. [After Y. S. Park, Phys. Rev. Letters, 13, 99, 1964.]

b. Exciton-Phonon Bound Particle

Another possibility of exciton–phonon interaction has been proposed by Toyozawa [40,41]. According to him, exciton–phonon bound states or quasi-bound particles could be formed, particularly in ionic semiconductors [40]. Theoretical calculations show that because of the internal binding energy of the complex, the energy of the first bound state is approximately reduced by 10%.

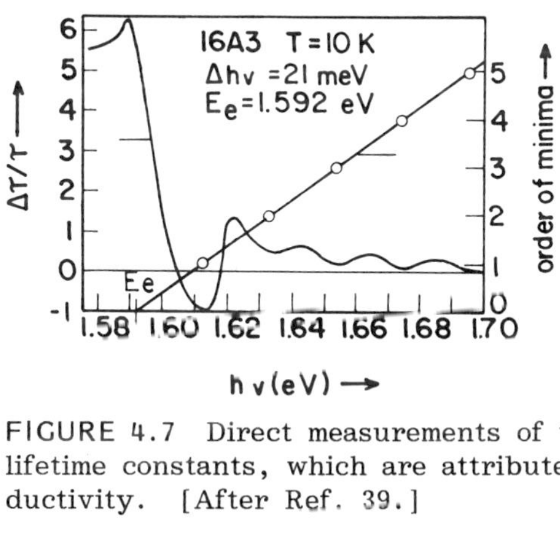

FIGURE 4.7 Direct measurements of the variations in mobilities and lifetime constants, which are attributed to oscillations in photoconductivity. [After Ref. 39.]

In other words, the separation between the exciton line and the phonon peak is reduced by 10%. The energy of the first bound state is given by [40]

$$h\nu = E_{exciton} - (1 - \Delta E)\hbar\omega_0 \tag{4.27}$$

where the estimated value ΔE is 10% of the phonon energy.

In order to observe the exciton–phonon–quasi-bound particle the following conditions should be met.

1. Electron–hole binding energy must be small compared to the LO phonon energy.
2. Effective hole mass must be greater than the effective electron mass.
3. The coupling between the exciton and the LO phonon must be strong.

All these criteria are fulfilled in a majority of ionic materials, and therefore they are suitable for examining the exciton–phonon–quasi-bound particle (EPQBP). In fact, in some complexes, for example TlCl,TlBr the presence of EPQBP has been observed [42–44]. A reasonable amount of experimental data come from optical absorption spectroscopy [11,13,14], but only a few from photoconductivity technique. This could be due to the experimental difficulties in the latter. There are only two compound semiconductors, GaSe [45] and CdTe [46], in which EPQBP has been observed by photoconductivity so far. The binding energy of this quasiparticle is found to be 3.1 meV and 2.3 meV in GaSe and CdTe, respectively.

Figure 4.8 shows the photoconductivity spectrum of CdTe recorded at 8 K. A study of the temperature dependence of this peak was also carried out [46]. It was reported that at high temperature the peak becomes broader and at 38 K the quasi-particle dissociates. In fact, considering the binding energy, it could be expected to dissociate at a slightly lower temperature (28 K). The cause for this difference has not been explained.

From the above discussion it is clear that the photoconductivity spectrum does reveal the contribution from the impurity states and the exciton structure on the low-energy side of the main photoresponse, corresponding to the principal absorption edge. However, it is worthwhile pointing out that the contribution from the impurity states, in many cases, is much more predominant than the exciton structure, and hence only the former is used for radiation detection.

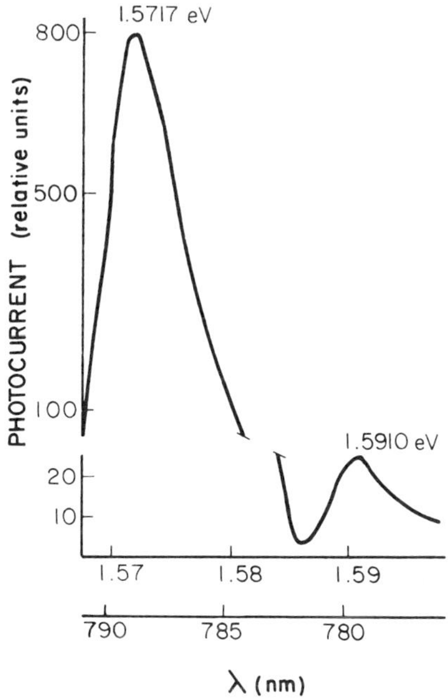

FIGURE 4.8 A peak corresponding to the exciton—phonon—quasi-bound particle. [After Ref. 46.]

4.2.5. Photoconductivity from Deep Valence Band Structure

It is not customary to examine photoconductivity on the higher-energy side of the band gap, E_g, because the radiation with higher energy than the band gap is mainly absorbed at the surface, where the surface recombination rate is higher and therefore the contribution in photoconductivity is much less. Moreover, the electronic transitions from the bottom of the valence band to the higher part of the conduction band are restricted by a selection rule based on the symmetry point in the Brillouin zone [5]. For both reasons, the spectral response curve generally shows a smooth decrease after the peak response, corresponding to the principal absorption edge.

Recent experimental studies on photoconductivity [47—49] show that in some cases, the valence band structure does extend the photoresponse curve on the high-energy side by revealing structural details. It is therefore necessary to know more about electronic structure very close to the top of the valence band.

a. Valence Band Splitting

The valence band is generally derived from p states of the constituent elements, and the energy depends not only on the angular momentum but also on the direction of spin of the electrons. Theoretical studies of spin-orbital interactions have been carried out extensively [50], and now it is known that the interaction of directed spin with the orbital angular momentum causes the valence band to be split. This can be understood very easily by considering group theory [51,52]. The spin effect is included in the band structure calculation by writing the Hamiltonian [53]

$$H\psi = \left[\frac{P^2}{2m} + V + \frac{h}{4m^2c^2}\hat{\sigma} X \nabla v \cdot P\right]\psi \tag{4.28}$$

where σ is Pauli's spin operator and is given by 2 X 2 spin matrices. The new Hamiltonian obtained by including spin-orbit interaction is invariant under the transformations applied to both functions (orbital and spin). The spin function changes sign under a rotation of 360°, and the number of operations of the symmetry group is doubled. Of course, new irreducible representations are needed to describe the properties of this new "double group." It is beyond the scope of this treatment to discuss characteristic properties of the double groups and the compatible relations between the irreducible representations of single and double groups. It is sufficient to understand here that at Γ point symmetry, threefold-degenerate p levels (six levels including spin) are split into a twofold level and a fourfold level by spin-orbit interaction.

Even though it provides information about the wave functions and the splitting of the energy states, group theory cannot help us estimate the magnitude of the separation, which can be obtained by using the perturbation method [54]. The theoretical procedure that is widely used, known as the $\underline{k} \cdot \underline{p}$ method, yields very satisfactory results. It is probably the single most important theoretical method for calculating the magnitude of the splitting of the valence band structure, provided that the effective masses at $\underline{k} = 0$ and the band-gap value are known. In practice, these parameters, and therefore the splittings, are known. For completeness and evaluation of the magnitudes of the separations, theoretically obtained equations are given below. It is worth mentioning that in this case the energy is measured with respect to the top of the valence band and not the bottom of the conduction band.

$$E_{hhv} = \frac{\hbar^2 \underline{k}^2}{2m^*}$$

$$E_{s0} = -\nabla + \frac{\hbar^2 \underline{k}^2}{2m^*} - \frac{\underline{k}^2 \underline{p}^2}{3(E_g + \Delta)} \qquad (4.29)$$

$$E_{lhv} = \frac{\hbar^2 \underline{k}^2}{2m^*} - \frac{2\underline{k}^2 \underline{p}^2}{3E_g}$$

Δ is the separation due to spin-orbit splitting and E_{hhv} and E_{lhv} are the heavy-hole and light-hole valence band, respectively. The band shape and the separation estimated from the above equation are shown in Fig. 4.9.

The magnitude of the splitting in II-IV and III-V compounds and in some of the ternary compounds could be of the order of 0.5 eV or sometimes even higher. The transitions from the lower part of the valence band to the top of the conduction band extend the photoresponse toward the high-energy side of the band gap.

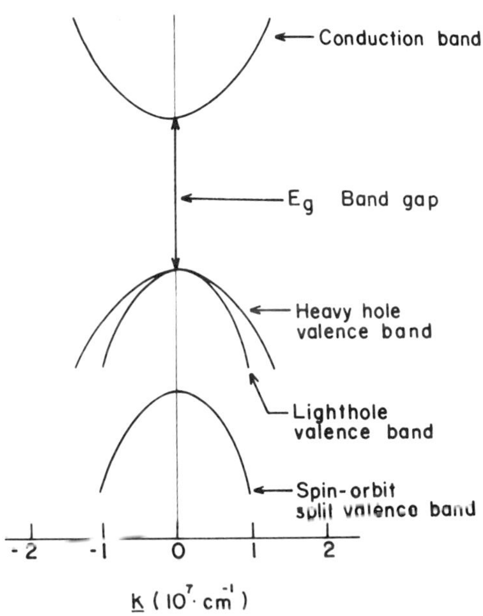

FIGURE 4.9 E—$\underline{k}$ curves obtained on the basis of the simple band model. Spin-orbit splitting of the valence band is exaggerated.

b. Crystal Field Splitting

Energy states in the valence band are further split by the crystal field which the electron sees. The origin of the crystal field splitting, first presented by Bethe [55], is very easy to understand but not so easy to calculate. We know that electrons with the same principal (n) and angular (l) quantum numbers belong to the same irreducible representation of the rotational group; therefore, these energy levels are degenerate. For example, in silicon, where the electron distribution is given by $1s^2 2s^2 p^6 3s^2 p^2$, the energy of $2p^2$ electrons is the same. If the atom is placed in a crystal lattice, then electrons see a potential field created by the crystal, which is no longer of spherical symmetry, and hence the degeneracy is partially or totally lifted. Group theory, as usual, provides qualitative information about the splitting; that is to say, we can conclude (assuming that the site symmetry is known) that the three degenerate p states will be decomposed into one nondegenerate level and one doubly degenerate level or three nondegenerate levels according to site symmetry. The magnitude cannot be estimated from group theory and here also the perturbation technique must be employed. The study of the ordering of the levels and their separation under certain symmetrical environments is classified as "ligand field theory," and the details are given in several places (see, e.g., Ref. 17). Our interest here is to know that the valence band is split according to the symmetry of the crystal, and the energy states have a certain symmetry; obviously, the transitions originating from these states have polarization. The effect of crystal field, T_d, and spin-orbit interaction on the valence band is shown in Fig. 4.10. It can be seen that the degeneracy of both p and d orbitals is partially lifted. These states participate in the photoconduction process, extending the photoresponse on the higher-energy side.

Such an effect has been predominantly observed in copper-based ternary semiconducting compounds, which are analogs of II-VI or III-V compounds. The upper part of the valence band of these compounds (e.g., a family of Cu-III-VI_2) are strongly influenced by 3d levels of copper [56—60]. An additional structure in the valence band is observed due to 3d levels. The fivefold degeneracy of wave functions is completely lifted due to spin-orbit interaction and the tetragonal crystal field. Shay et al. [60] also obtained direct experimental confirmation for d-level splitting through the electroreflection spectrum of $CuInSe_2$. The symmetry of the levels has been verified by examining the polarization dependence [6,61]. On the basis of theoretical and experimental investigations, it has been concluded that d levels are split in tetrahedral symmetry into a threefold Γ_{15} and a twofold Γ_{12}. Further, taking into account spin-orbit interactions, Γ_{15} is split in Γ_7 and Γ_8. The ordering of

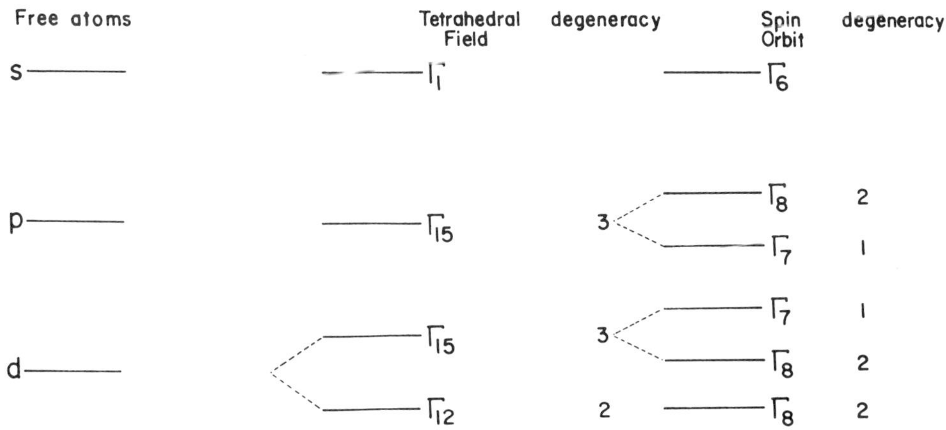

FIGURE 4.10 A systematic diagram showing the splitting of the atomic levels in T_d symmetry. Splitting due to spin-orbit interaction is also shown along with the degeneracy of the respective states.

the states is shown in Fig. 4.10. The complete analysis of these aspects has been presented by Shay et al. [60]. It has been suggested that d levels of copper have a strong influence on the valence band structure of Cu-III-VI_2 compounds. Similar arguments are valid for Ag-III-VI_2 compounds.

In fact, the assumption of a pure d wave function of copper in copper-based ternary compounds (Cu-III-VI_2) is an oversimplification. The substantial contribution from the d orbitals of copper stems from the fact that the energies of these orbitals are very close to those of the p orbitals of group VI elements. Obviously, there exists strong hybridization of 3 d wave functions of copper and p wave functions of the elements of group VI. Naturally, the character of the final wave function depends upon the contribution of d and p wave functions; an extensive study in this direction has been carried out by Jaffe and Zunger [56,57] and band structure calculations have been done [56–58] by using the recently developed potential variation mixed basis approach. This technique is more useful, as it permits us to solve the many-electron problem self-consistently [59]. Because of these efforts, several properties such as electronic charge distribution, chemical bonding, and density of states are now understood. It is clear from the theoretical calculations of band structure [56–58] that the upper valence band of these compounds show a well-pronounced density of states, originating from the interaction of 3d states of copper with p orbitals of the elements of group VI. Copper d characters are dominant in the upper part of the valence band. Figure 4.11 shows an approximate scheme of a valence band altered by the presence of 3d states of copper in $CuInSe_2$.

Experimental results on the electronic band structure also confirm the presence of a high density of states near the top of the valence band and show explicitly the contribution of 3d states of copper and p states of the group VI elements. The presence of high-density states in the upper part of the valence band and their d wave functions have important consequences with respect to the transition probability and the form of the optical absorption edge.

In this case, the treatment given in Section 4.2 is not expected to be fulfilled, as neither density of states nor transition probability follows the assumptions employed therein. Care should be taken to use Eq. (4.21) for photoconductors where the hybridization of orbitals is strong. Moreover, the upper part of the valence band (not just the top of it) also contributes in optical transitions. Naturally, the photoconductivity spectrum is extended toward the high-energy

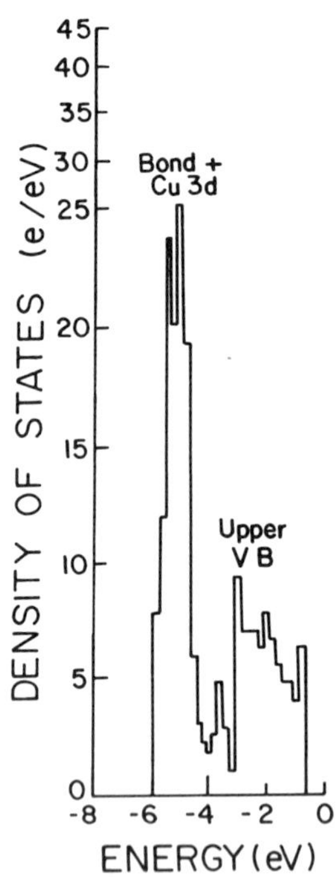

FIGURE 4.11 A typical density-of-states curve for $CuInSe_2$. [After Ref. 56.]

side. This is confirmed not only in the 3d states of copper but also in transition metals such as manganese.

With this view, the response curve of photoconductivity spectra should be examined toward the high-energy side of the band gap. Recently, such a study was carried out for a single crystal of $AgInS_2$ that was grown by chemical transport and others grown by Bridgman method [47]. It has been confirmed that the contribution from d levels does take place and, indeed, the spectral response is extended considerably toward the high-energy side.

The calculated band structure has been confirmed by absorption spectroscopy, reflectivity, and electroreflectivity data [6]. As far as photoconductivity is concerned, there are two major effects.

1. The photoresponse up to a certain limit is extended toward the high-energy side in copper-based ternary compounds. The extension and the form of the spectrum depend upon the copper-centered density of states in the valence band.
2. A high density of states in the valence band can create an unexpected effect—a photoquenching—in p-type semiconductors. This is because the electrons from these states are ejected to the top of the valence band and recombine to the holes, and therefore photoquenching is observed. This phenomenon has been reported by Joshi and Echeverria [61] in $CuGaTe_2$. The details will be discussed in Chapter 7.

Another important aspect of the valence band structure that has not received due consideration in photoconductivity studies is the perturbation caused by the adsorption of atoms (both chemisorption and physisorption). It is known that almost all elements can be adsorbed on the surface of the photoconductor. The effects of the transition metals are probably strongest because the wave functions of d orbitals are directional.

Experimental investigations on semiconductor surfaces have been revolutionized in the last decade, and now a great deal of evidence is available [62] for the presence of additional levels of electronic structure in the valence band due to the adsorption complex. There are several models to explain this, but discussion of them lies outside the scope of this volume. What is clear is that the valence band contains electronic structure that varies with the adsorbed particle and the crystallographic plane of the surface.

The contribution of these states in photoconductivity is generally neglected. Recently, photoquenching originating from oxygen adsorbate surface states has also been observed by Echeverria et al. [63]. With the presence of these states, the previously observed but never explained behavior in p-type GaAs—namely, shifting of the photoconductivity peak on the low-energy side with respect to the absorption peak [63]—is explained satisfactorily.

In the above-mentioned cases, the valence band structure shows additional states, and in specific situations they can lead to photoquenching or even negative photoconductivity. The details of these phenomena will be discussed in Chapter 7.

4.2.6. Diluted Magnetic Semiconductors

Very recently, a novel class of semiconductors—semimagnetic or diluted magnetic semiconductors—have attracted great interest because of their theoretical properties and potential technological applications. Komarov et al. [65] reported for the first time a giant enhancement in the magneto-optical effect. Since then, several dramatic properties (such as giant negative magnetoresistance, spin glasses with short-range interaction, magnetic polaron, substantial Faraday rotation, and paramagnetic spin glass interaction) of significant importance have been examined, and encouraging results have been reported [66,67].

One of the well-known diluted magnetic semiconductor (DMS) systems is a ternary alloy of II-VI compounds formed by introducing a transition metal as a third constituent element (not as an impurity). The special interest in transition metals is generally connected with their ability to form compounds and alloys in which the outermost set of stable d electron orbitals is partially filled. It is the interaction of incomplete d electrons that is responsible directly or indirectly for the optical and magnetic properties. If these features are added to conventional semiconductors such as II-VI compounds, additional properties unknown to semiconductors could be expected. Moreover, transition elements exhibit an unusual "variable valency phenomenon" that extends their capacity to form compounds or alloys.

The replacement of an element of group II of the periodic table by an element, such as manganese, of group VII A would seem in many ways to be a misfit in ternary alloys based on II-VI compounds and should create a large number of many kinds of defects. Even the solubility of a high percentage of manganese in a solid solution of II-VI compounds is not so obvious, since manganese and cadmium (or zinc or mercury) are chemically very different. However, a close look shows that there is a hidden similarity—both have two electrons in the outermost orbit with s wave functions, and the previous orbit is stable. In the case of a group II element, the last but one orbit is completely filled up, while in the case of manganese the 3d orbital is exactly half-filled, but both of them are stable against removing or adding an extra electron in the system. It is worth mentioning that the solubility of the iron atom, which has unstable 3d orbitals, is scarcely 2—3% in II-VI tellurium compounds [67]. In addition to this, the homogeneity of an iron-based diluted magnetic semiconductor is very poor. For both these reasons, these materials are less investigated than manganese-based diluted magnetic semiconductors.

The electrical, optical, and magnetic properties of several alloys such as $Mn_xCd_{1-x}Te$, $Mn_xCd_{1-x}Se$, and $Mn_xZn_{1-x}S$ have been examined, and their usefulness for devices has been realized. The value of x up to which a homogeneous solid solution can be formed for different II-VI compounds for manganese metal can be understood with the help of Fig. 4.12. The crystalline structure for the corresponding concentration of the transition metal is also shown in Fig. 4.12.*

As mentioned earlier, DMS alloys are formed by substituting the transition metal in the site of the group II element. The outer electronic structure of transition elements is given by the $3d^5\ 4s^2$ configuration. Thus, $4s^2$ electrons (as in the cadmium atom) take part in the binding process. The 3d orbitals are more localized near the transition metal site. This situation is totally different for 3d orbitals of copper in copper-based ternary compounds, where they are hybridized with p orbitals. Several theoretical and experimental results confirm that 3d orbitals maintain their character. The energy states are shifted and degeneracy is lifted according to the site symmetry of the transition metal ion, which leads to the spread of atomic levels over the band-gap region. The ground state lies in the upper part of the valence band or just at the top of it. The excited states lie within the band gap and in the conduction band. The precise relative positions with respect to the conduction or valence bands depend upon the material and are determined experimentally.

Here we will limit ourselves to the effect of the presence of transition metal ions on the valence band structure and its consequences in optical absorption and photoconductivity measurements.

---

*I understand that very recently good quality crystals of new families of diluted magnetic semiconductors derived from II-VI compounds have been sucessfully grown. In these materials, manganese ions are replaced by other transition metals ions. Typical examples are $Co_xZn_{1-x}Se$ ($x \leq .20$), $Co_xZn_{1-x}S$ ($x \leq .20$), $Ti_xZn_{1-x}Se$ ($x \leq .05$), $Ti_xZn_{1-x}S$ ($x \leq 0.5$), $V_xZn_{1-x}Se$ ($x \leq .25$), $Cr_xZn_{1-x}Se$ ($x \leq 0.20$), $Zn_xNi_{1-x}Se$ and $Zn_xNi_{1-x}S$. Unfortunately, the experimental investigation is incomplete and it is not possible to report the diagram as shown in Fig. 4.12 for Mn. These DMS show several interesting properties, the details of which are expected to be published soon. (Private communication from Dr. W. Giriat, Instituto Venezolano de Investigacion Cientifica, Caracas, Venezuela.)

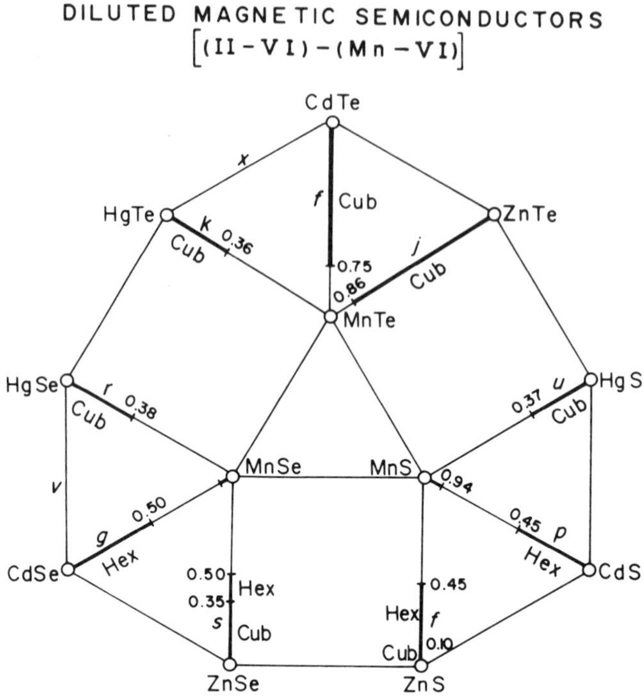

$Mn_j Zn_{1-j} Te$ $Mn_f Cd_{1-f} Te$ $Mn_k Hg_{1-k} Te$

$Mn_s Zn_{1-s} Se$ $Mn_g Cd_{1-g} Se$ $Mn_r Hg_{1-r} Se$

$Mn_t Zn_{1-t} S$ $Mn_p Cd_{1-p} S$ $Mn_u Hg_{1-u} S$

FIGURE 4.12 Manganese-based diluted magnetic semiconductors derived from II-VI semiconducting compounds. The figure shows the concentration of manganese and the crystalline structures.

It is well established that when the transition metal ions are introduced in the matrix of II-VI semiconducting compounds the band structure is affected (even without a magnetic field) in several ways:

1. The band gap changes according to the concentration of the transition metal.
2. 3d orbitals and higher excited states (e.g., 4G of Mn) modified by the ligand field and the position of the upper levels are pinned with respect to the conduction or valence band.

3. A comprehensive understanding of the effects of the transition metals in the valence band structure has not been achieved, and the precise role of 3d orbitals is not clear. However, experimental investigation of $Cd_{1-x}Mn_xSe$ show [58] that 3d orbitals contribute in two ways. First, they widen the valence band, and second, higher density of states is created in the lower part of the valence band.
4. The absorption both from d-d levels and from the valence band (or energy states close to it) to the conduction band takes place in the same spectral range. In certain symmetries, for example, tetrahedral, the intensities of d-d transitions are substantially high. Naturally, peaks corresponding to these transitions are superimposed on the main absorption edge. The positions of these peaks are independent of the temperature and concentration of the transition metals, and therefore this fact serves as a guide to identify d-d transitions.

Considering the importance of d levels in absorption and electro-optical properties, it is necessary to understand the origin and magnitude of the splittings and the order in which they are spread. This can be done by considering the form of 3d wave functions and their angular dependence, which represents the charge density of the electrons. This information is obtained from elementary quantum mechanics and is given in Fig. 4.13.

A simple way to visualize the effects of the crystal symmetry on 3d orbitals is to study the transition metal ions in two different surroundings: at the center of the cube and at the center of the tetrahedron. The latter is directly relevant to the present problems.

Let us consider, in the first place, a transition metal ion placed at the center of a cube of side "a" with a point negative charge located at the center of each face. It is clear from Fig. 4.14 (without considering group theory) that d_z^2 is repelled by the negative charges placed at 00a, while the orbitals d_{xy}, d_{yz}, and d_{zx} are relatively slightly affected as their charge distributions are in the xy plane, symmetrical to the x and y axes. Consequently, electrons will try to avoid d_{z^2} and $d_{x^2-y^2}$ orbitals and try to be in a d_{xy}, d_{yz}, or d_{zx} orbital (i.e., these orbitals have lower energy than the others). This is not true if the ion is in a tetrahedral symmetry (see Fig. 4.14). The d_{xy}, d_{yz}, and d_{zx} orbitals are nearer to the ligand atom than the $d_z{}^2$ orbital, and hence, in this symmetry, electrons would rather be in d_{z^2}, $d_{x^2-y^2}$ orbitals (i.e., in tetrahedral, T_d, symmetry, d_{z^2} and $d_{x^2-y^2}$ orbitals have lower energy). Splittings of d orbitals in these two symmetries are shown in Fig. 4.14.

The d-orbital energy-level splitting is measured in units of Dq and is defined as the difference between d_γ (a set of d_{z^2} and

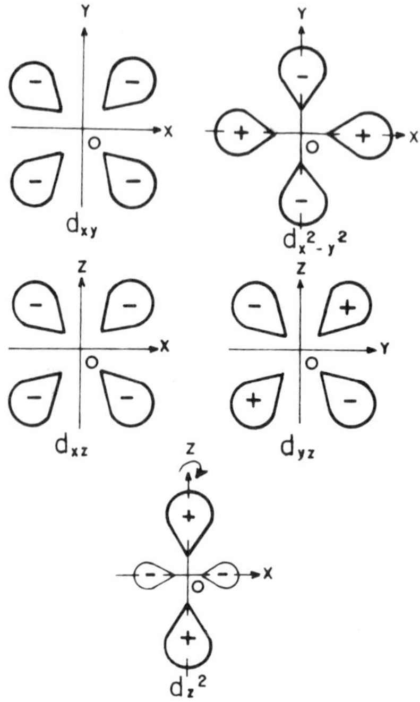

FIGURE 4.13 The shape of d orbitals in spherical symmetry.

d_{x2-y2} orbitals) and d_ε (a set of d_{xy}, d_{yz}, d_{zx}) levels which is 10 Dq. The magnitude of Dq is proportional to the strength of the crystal field generated by the ligands [17]. This in turn depends upon, among other factors, the charge and size of the ligand, its polarizability.

The average energy of the ion is shifted by the interaction of negative charges; however, such a change is not shown as it is the usual practice to give the effect as simply the splitting of the various d orbitals relative to each other. Using group theory, the splitting of energy levels in a particular symmetry can be worked out. Details of calculations and tables are available in the literature [16].

Over the last few years experimental investigations on semimagnetic compounds have revealed that alloys of MnTe and CdTe have received great attention and much data are now available. We will therefore discuss the effect of manganese on the band structure of CdTe (i.e., the system $Cd_{1-x}Mn_xTe$). However, similar arguments can be extended to alloys of other II-VI compounds.

Manganese atoms have five electrons in 3d orbitals. Using Russell-Saunders coupling, the ground state of manganese is found

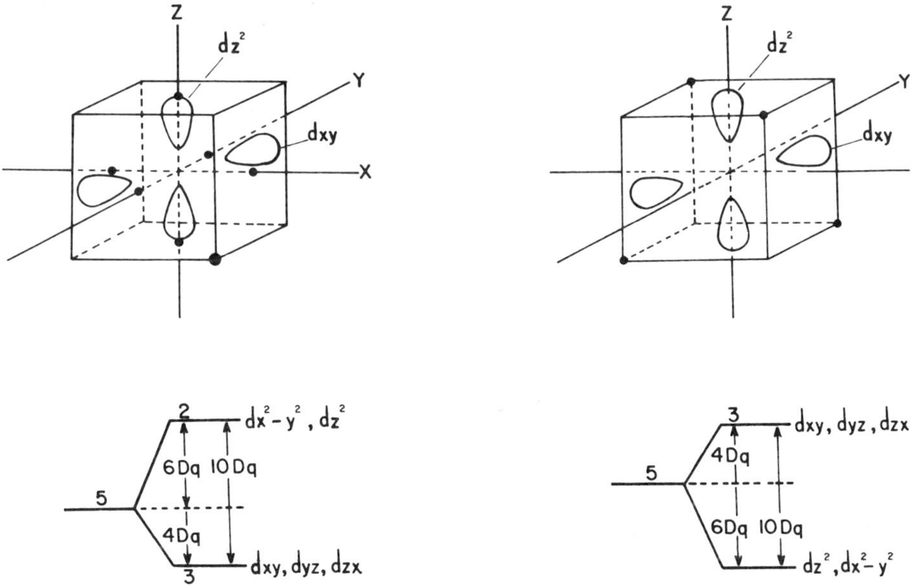

FIGURE 4.14 Splitting of d orbitals in octahedral and tetrahedral symmetry. [From Ref. 17. Reprinted by permission of John Wiley & Sons, copyright, 1966.]

to be 6S and the first excited state is 4G. Using the group theoretical analysis for T_d symmetry, it can be seen that the excited state is split into five states, the ordering of which is invariable; what varies is the magnitude of the splitting from compound to compound. As there are plenty of data available for the $Cd_{1-x}Mn_xTe$ system, it is possible to represent the position of energy levels with respect to the conduction and valence bands (Fig. 4.15).

To avoid confusion, we have kept the standard nomenclature for these states. This notation tells how the wave functions transform under the symmetry operation T_d. The consideration of symmetry is very important for determining the transitions between two states. This aspect will not be discussed here, but the transitions $4T_1-6A$ and $4T_2-6A$ are allowed. These transitions are atomic in origin, and a contribution from them is not expected in the photoconduction process. However, in certain circumstances these transitions do contribute, and this will be discussed shortly.

The precise estimation of the separation between the energy states of the d system in the presence of crystal fields and the interelectronic repulsion is really a complex issue. It is not possible to summarize the general methods involved in such calculations here,

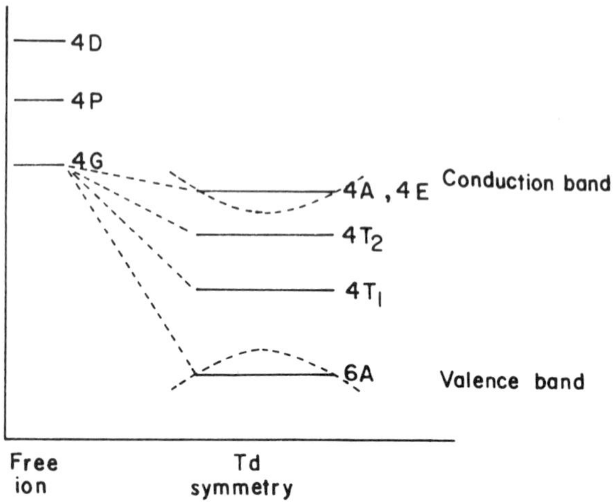

FIGURE 4.15 The splitting of the ground state of the $3d^5$ level of Mn in T_d symmetry. The relative position of the valence band depends on the alloy.

but the results are generally presented in the form of Tanabe-Sugano diagrams. A typical diagram for manganese in T_d symmetry is shown in Fig. 4.16. This diagram is valid only for a given ratio of two parameters $\underline{C}/\underline{B}$, both of them related to interelectronic repulsive energy. It is not possible to predict how a change in $\underline{C}/\underline{B}$ alters the Tanabe-Sugano diagram, but a separate calculation has to be carried out for a specific value of $\underline{C}/\underline{B}$. Similar diagrams for other symmetries are also available. With the help of this diagram and with knowledge of the position of a few lines, important parameters such as strength of the crystal field, the interelectronic repulsion parameter, and the position of higher states can be roughly obtained. Up to now, no theoretical work has been carried out to locate the relative position of the ground state and higher excited states with respect to the conduction or valence bands. This information is obtained through combined experimental investigation on optical absorption and photoluminescence.

As mentioned earlier, d—d transitions, even though they originate from localized states, participate in the photoconduction process. There is experimental evidence [69—72] for crystal field—induced photoconductivity in ternary compounds and DMS compounds; however, there is no well-established theory to explain these observations. Several possible mechanisms for photoconductivity caused by crystal field absorption (d—d transition of 3d configuration of transition metal ion) are given below.

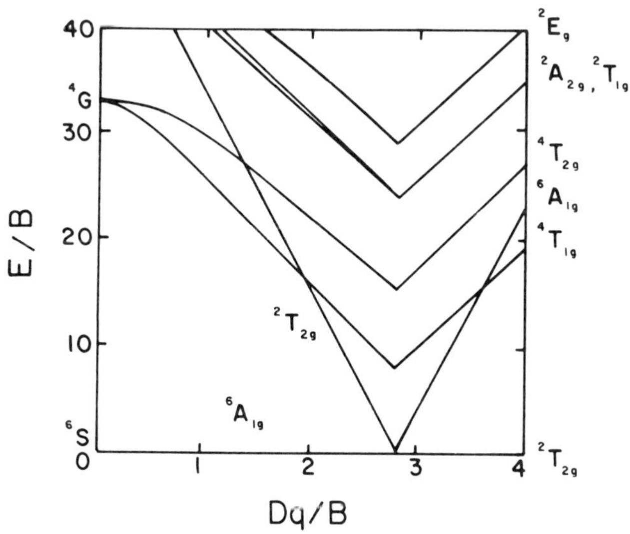

FIGURE 4.16 Tanabe–Sugano diagram for $3d^5$ electrons in T_d symmetry. Energy separations for other electron configurations and different symmetries are available. (Valid for C/B = 4.48.) [From Ref. 17. Reprinted by permission of John Wiley & Sons, Inc. Copyright 1966.]

1. When the excited state lies very near the bottom of the conduction band, then it is possible that its wave function mixes with the wave function of the conduction band (plane wave function) [70] and photoconductivity will be observed. A typical example is photoconductivity in ZnO due to the d–d transition of the cobalt ion [70].
2. Photoconductivity and photoluminescence studies in $CdIn_2S_4$:Cr prove that crystal field–induced energy states of transition metal ions and one electron energy band structure are not isolated from each other [72]. There is a transfer of energy between these two systems. In this case, by artificially inducing an energy state into the band-gap region, the energy transfer mechanism has been visualized. In an interconnected system, several alternative schemes for the transfer of electrons from excited states to the conduction band can be worked out. However, it is difficult to obtain direct experimental confirmation for any of them.
3. Luminescence originated from the excited state of the transition metal ion might take place, and the electron from the ground state of the ion can be transferred to a nonionic state [70], that is, to a crystalline system.

4. Recently, an alternative probable explanation was proposed by Martin et al. [49]. According to them, the excited states form a resonant state in the conduction band. The probability of an electron decaying in the conduction band is much higher than in the valence band, and therefore the electron becomes part of the conduction band, losing its identity with the transition metal ion.

Whatever may be the mechanism for crystal field induced photoconduction, we know from various experimental results that in several circumstances d—d transitions do contribute in the photoconduction process. This has an important consequence. It is possible to select a wide-band-gap diluted semimagnetic semiconductor with a proper transition metal and, by adjusting its concentration, pin the position of crystal field split 3d levels (upper and excited) with respect to the position of the valence and conduction bands. The system designed in this manner has the potential to be a high sensitivity detector in the near-ultraviolet region. A few attempts have also been made in this direction.

The purpose of the present discussion is to bring to notice another energy scheme that could be more effective (in a certain energy range) for photodetection purposes than the conventional one, in which band-to-band transitions are used to create mobile charges. We believe that the former system has some inherent advantages for improving the sensitivity.

Let us suppose that the crystal field split 3d energy states causing absorption and hence photoconductivity have a separation less than the band-gap value. In this case, the radiation will penetrate deep in the semiconductor and the major contribution in the absorption will take place in the bulk of the semiconductor instead of at or near the surface, where the surface recombination velocity is very high. Naturally, the mobile charge carriers produced by absorption are used effectively in the conduction process. In addition to this, there would be a greater number of absorbed photons because the reflectivity is lower for photons with energy less than the band gap.

In this case, the number of photons absorbed depends on the thickness of the semiconductor, and therefore the quantum efficiency must be defined as a function of thickness. From the definition given in Chapter 1 and including the effect of the reflectivity, the quantum efficiency is given by [49]

$$\eta = [1 - R(\lambda)]\,\alpha \int_0^t e^{-\alpha x}\,dx \qquad (4.29)$$

where x is the distance perpendicular to the plane of incident radiation. In spite of the above advantages, unfortunately, sufficient efforts have not been made from the technological point of view to exploit the potential benefits these materials may provide.

At the beginning of the development of the DMS system, it was thought that even though the alloys formed are homogeneous, the presence of transition metal ions in II-VI compounds could create defects and strain and therefore these alloys would not be suitable for the fabrication of devices. Nevertheless, recently it has been confirmed that the quality of the alloy is excellent for devices [69]. These materials have a bright future.

4.3. SUPERLATTICE

Superlattices are a new type of material, not from the point of view of chemical composition but from the point of view of optical, electrical, electro-optical, and transport properties. They are very important materials for ultrahigh-speed photodetectors [see Chapter 6] and other photonic devices. Band structure, and therefore optical and electro-optical properties of the superlattice structure, are remarkably different from those of conventional materials and hence deserve special consideration.

In 1969–1970, Esaki and Tsu [73] proposed a one-dimensional periodic structure of alternating ultrathin layers of semiconductors with a period less than the electron mean free path. This system, known as a superlattice structure, is shown in Fig. 4.17. In this case, the entire electron system enters into a quantum regime exhibiting new phenomena that are completely unknown in the usual three-dimensional solids. Since then, extensive work in theoretical, experimental, and technological fields has been carried out [74–85], including studies in optical absorption [75,76], photoluminescence [77], magnetoquantum effects [78], Raman scattering [79], quantized Hall effect [80], and modulation doped high-speed devices [81–

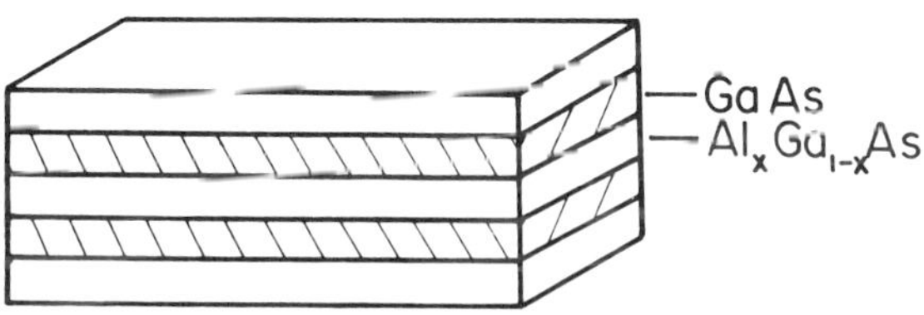

FIGURE 4.17 A widely used superlattice structure of the $Al_xGa_{1-x}As$-GaAs system. The layer thickness varies from ~30 Å to ~250 Å.

84]. For our present purpose, we will limit ourselves to optical absorption and photoconductivity [85].

Optical absorption depends upon the density of states in the conduction and valence bands. The superlattice structure alters the potential, and hence the band structure of the individual materials. The variation in the structure depends upon the periodicity of the superlattice, which is much greater than the original lattice constant. This means that the Brillouin zone is divided into a series of subzones that form allowed subbands separated by forbidden regions. Figure 4.18 shows the density of states for electrons in a superlattice. It is quite different from conventional three-dimensional solids. In the former case, the density of states varies like a staircase structure, while in the latter case it varies in a parabolic manner. These two forms are also shown in Fig. 4.18 for comparison. It is clear that the density of states and the optical absorption coefficient vary rather smoothly near the band-gap region.

Since a superlattice is formed by alternate layers of different semiconductors and the band gap is different for each, the band structure of the superlattice cannot be constant in space, but a spatial variation is observed in the positions of the conduction and valence bands. Figure 4.19 shows spatial variation in the energy scheme, which has far-reaching consequences from the point of view of both basic physics and the technology of semiconducting devices.

The interesting feature of the absorption spectrum is that, unlike three-dimensional solids, rather sharp lines are superimposed on the band-edge transitions. These transitions correspond to the

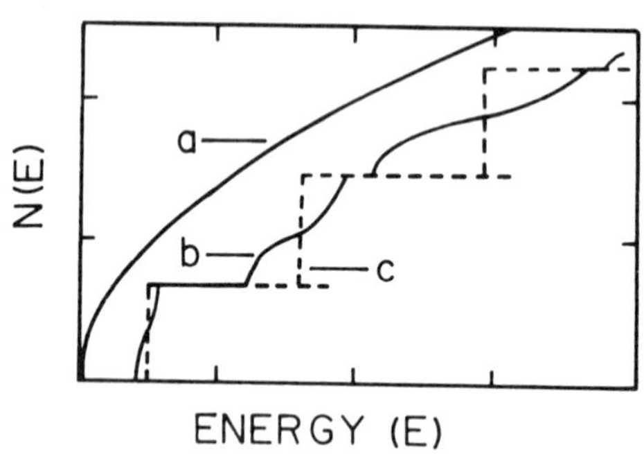

FIGURE 4.18 Density of states as a function of energy. (a) Three-dimensional solid; (b) superlattice structure; (c) two-dimensional solid. [From L. Esaki, A bird's eye view on the evolution of semiconductor superlattice and quantum wells. IEEE J. of Quantum Electronics 22; 1611, 1986.]

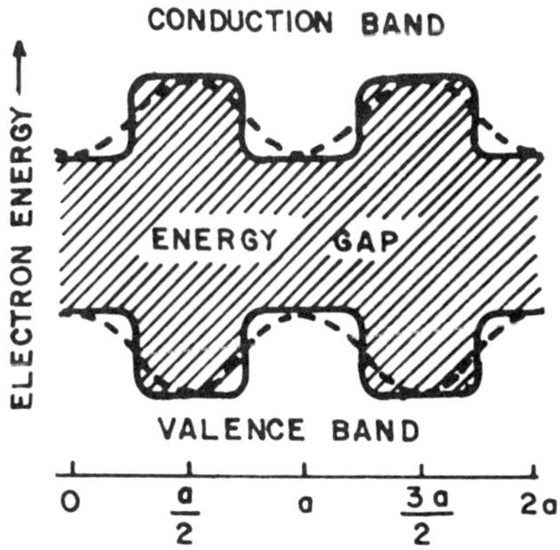

FIGURE 4.19 Spatial variation of conduction and valence band edges in compositional superlattice. "a" is the periodicity of the lattice.

bound states in isolated and coupled quantum wells. Convincing experimental evidence is reported by Dingle et al. [76] on the $Al_xGa_{1-x}As$- GaAs heterostructure. The optical absorption spectra at low temperatures clearly show peaks corresponding to the excitation of an electron from the n = 1 heavy and light mass valence band bound states to the n = 1 conduction band bound state. The experimental values are in excellent agreement with the theoretical values of energy states obtained for a rectangular quantum well whose dimensions are determined by the growth conditions of superlattice structure. Further, it has also been confirmed from experimental results that when the separation between two wells is reduced, then the electron and hole bound states are split by resonant coupling, through the potential penetrable barrier, giving rise to symmetrical and antisymmetrical states. These are special features that modify charge carrier generation and are unknown in conventional semiconductors.

From the point of view of photoconductivity in superlattices or quantum wells, band structure and hence optical absorption are not the only important properties but also the transport of the photogenerated charge carriers perpendicular to the potential wells. Three transport mechanisms have been proposed; they can be visualized with the help of Fig. 4.20.

Phonon-assisted tunneling. This mechanism is suggested for those cases where the charge carriers are strongly localized within the wells. The only possibility of transport is through barrier penetration, and the energy to accelerate this process is taken from phonons.

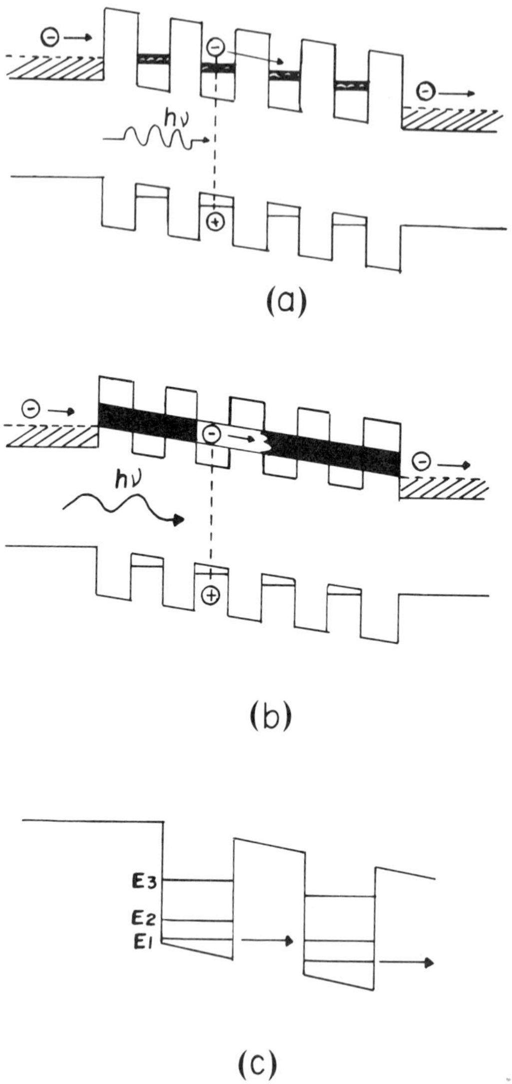

FIGURE 4.20 Band diagram of superlattice under bias voltage. Transport through: (a) phonon-assisted tunneling; (b) mini conduction band; (c) incoherent resonant tunneling. [After F. Capasso, K. Mohammed and A. L. Cho, Resonant tunneling through double barriers, perpendicular quantum transport phenomena in superlattices, and their applications. IEEE Jour. of Quantum Electronics QE-22, 1853, 1986.]

Transport through miniband. When the thickness of the potential well is small (comparable to the de Broglie wavelength), the wave functions of the electrons (or holes) of the adjacent well mix with each other due to tunneling. Thus, a sort of "miniband" is formed whose width is proportional to the probability of barrier penetration, and it contributes in the transport process [86].

Resonant tunneling. Transport in the quantum well structure has been observed and studied by both types of resonances, coherent and incoherent. The first type of process occurs when the energy and the lateral momentum of the occupied and unoccupied states in neighboring quantum wells are equal and the width of the barrier between them is small. Incoherent resonant tunneling takes place under a high electric field, where the situation is altered as the energy diagram is tilted (Fig. 4.20c). If the energy of the applied electric field is equal to the difference between the ground and excited states of the next quantum well, resonant tunneling will take place, the electron penetrates the potential barrier, and conduction is observed [87]. However, the voltage required for this process is relatively high (10^5 V/cm).

For these reasons, the effects of these quantum states in the photoconductivity spectrum have been investigated in this direction, and the results are reported by Tsu et al. [85]. This study is important because it correlates the band structure of superlattice quantum states originating from potential wells and anomalous transport properties.

The experimental confirmation for the above mechanism has been reported by Tsu et al. and is given in Fig. 4.21. When the separation between the wells is reduced, an increase in photoresponse is observed (at least a few times). This is because the probability of the superposition of the wave functions of adjacent wells increases and transport becomes easier. In this design, photoconductivity takes place through a tunneling process as demonstrated by Capasso et al. [86,87].

The other notable feature of the photoconductivity that originates through tunneling is that the contribution of the holes is eliminated. This happens because the probability of the tunneling potential barrier created by superlattice layers increases exponentially as the effective mass of the particle decreases. The effective mass of the electron is much less than that of the holes, and therefore, the electrons can penetrate the barrier; meanwhile, the holes remain localized in the quantum well. This effect is called the "mass filtering effect."

This means that electron mobility and hence the transit time of electrons perpendicular to the layers depend exponentially on the

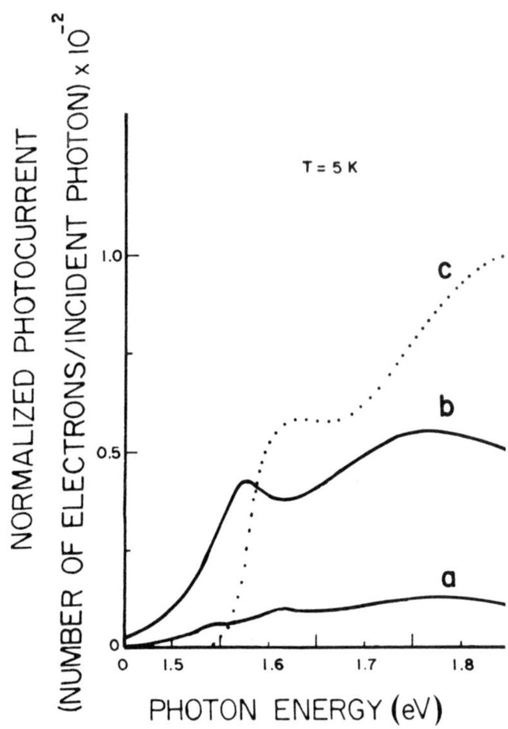

FIGURE 4.21 Normalized photoconductivity spectrum for $Al_xGa_{1-x}As$—GaAs system. The periodicity of layer thicknesses of each material was (a) 110 Å; (b) 50 Å; (c) 35 Å. [After Ref. 85.]

superlattice barrier thickness. Thus, photoconductivity gain and gain-bandwidth product can be tailored by adjusting the parameters of the superlattice. Once again, the validity of the arguments has been confirmed by fabricating a device with superior performance. High-gain, low-noise photoconductors whose operational capability can be extended to lower voltages are now possible. A device based on quantum-mechanical tunneling in the superlattice structure has been recently demonstrated. Excellent detectivity (8×10^{-14} W/$Hz^{1/2}$) has been reported [88].

Another curious aspect of photoconductivity in a superlattice is its voltage dependence. Because of the negative differential conductance, a peculiar property of the superlattice [89], a linear relation between voltage and photocurrent is not expected. Experimental data are needed to determine the voltage at which the photoresponse is maximum. Tsu et al. [85] have examined the voltage dependence of the photocurrent for the GaAs—$Al_xGa_{1-x}As$ superlattice. Figure 4.22 shows the experimental results reported by

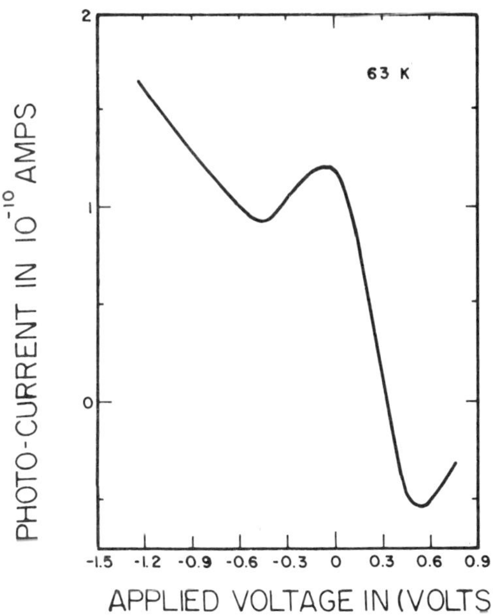

FIGURE 4.22 Dependence of photocurrent on applied voltage for superlattice structure ($h\nu$ = 1.55 eV and T = 65 K). [After Ref. 85.]

them. A positive voltage here means that the gold contact at the top surface has a positive potential with respect to the substrate. The figure shows a dramatic improvement in photoresponse with the electric field (negative bias). These results are particularly significant in improving the detectivity of the sensors. This point will be discussed further in Chapter 6.

4.4. MODULATION DOPED SEMICONDUCTOR HETEROJUNCTION SUPERLATTICE

The other outstanding application of the superlattice structure is modulation doped devices in which a high mobility for electrons is achieved [90,91]. A superlattice structure, because of its spatial variation in the band-gap edges, permits electrons to be separated spatially from their parents' impurities by doping them in the region of potential hills. This can be achieved by introducing impurities of the desired type (say n-type for donors) in material with a high energy gap. For example, in the $Al_xGa_{1-x}As$—GaAs system, donor impurity is introduced only in $Al_xGa_{1-x}As$. Electrons are separated from donor atoms by the band-gap potential and accumulate in the regime of GaAs. The density of electrons in this region is higher because they are confined by a high potential originating from $Al_xGa_{1-x}As$ on both sides. The discontinuity in the conduction band at the interface of doped $Al_xGa_{1-x}As$—GaAs is shown in Fig. 4.23.

The consequence of a spatial variation of band structure that physically separates mobile electrons from their parent impurity atoms is that high-density electron gas does not see the ionized impurity atoms (donors) that are responsible for the scattering mechanism. Obviously, the mobility in this case is expected to be higher than in GaAs or $Al_xGa_{1-x}As$ grown under the same experimental conditions. Dingle et al. [75] successfully demonstrated this concept in modulation doped GaAs—$Al_xGa_{1-x}As$ superlattices and confirmed that the mobility exceeds the Brooks—Herring predictions. The effect is more remarkable at low temperatures. At 77 K, the mobility is increased by nearly six times by modulated doping [92]. This is certainly a very significant quality, as the transit time of the electrons, which is an important parameter, is reduced and hence the device speed is increased considerably [93,94]. Using

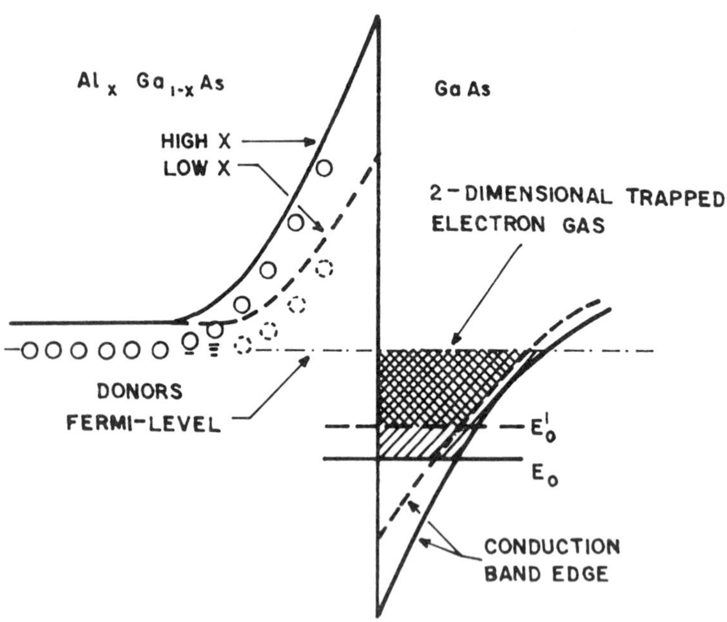

FIGURE 4.23 Conduction band discontinuity at $Al_xGa_{1-x}As$ (doped)—GaAs (undoped) interface. [After Ref. 94.]

this property, superlattice photodetectors have already been reported, and some of them will be referred to later.

In the modulation doped heterostructure, not only is the conduction band affected, but the valence band also shows alteration. Sooryakumar [95] observed a large band-gap mixing in the ground valence subband. This property of the band structure leads to a high emission and absorption rate that is just suitable for photonic devices.

A remarkably high mobility is an attractive property of this structure, and it has been used in many other devices such as high-speed FETs, high electron mobility transistors (HEMT), and two-dimensional electron gas field effect transistors (TDEGFETs). The details of these high-speed devices are not pertinent here. The role of the superlattice structure in photodetection is clearly established, and we will witness the high performance of these detectors in the coming years.

4.5. THE n-i-p-i SUPERSTRUCTURE

Another approach to changing the band structure periodically in space and to tailoring the electrical and optical properties is to dope donors and acceptors alternately in the same material. Recent technological developments in growth processes have experimentally proved that it is possible to control with precision both the doping concentration and the thicknesses over which it is carried out.

Initially, let us suppose that donors and acceptors are not ionized and that they are in an unstable condition. Depending upon the temperature, a large number of electrons from donors will be recombined with holes from acceptors. Thus, ionized atoms (donors and acceptors) create a region of space charge in the intrinsic semiconductor. A positive space charge of the magnitude eN_D^+ in the n region and a negative space charge of magnitude eN_A^- in the p region are created. This leads to a buildup potential whose magnitude depends upon the impurity concentration of donors and acceptors. Thus, a periodic and rather oscillating band structure is created whose period depends upon the period of doping, which is controlled experimentally. Naturally, the structure of the valence and conduction bands is modulated by the additional periodicity achieved by doping. Let us suppose that the superstructure period is about N times that of the lattice; then the valence and conduction bands of the pure crystal will split into N subbands. This feature is very important for optical properties and allows the transitions of higher energy than that of the band gap. Figure 4.24 shows the spatial variation of band structure and subband transitions whose capabilities are not exploited fully from the point of view of device technology.

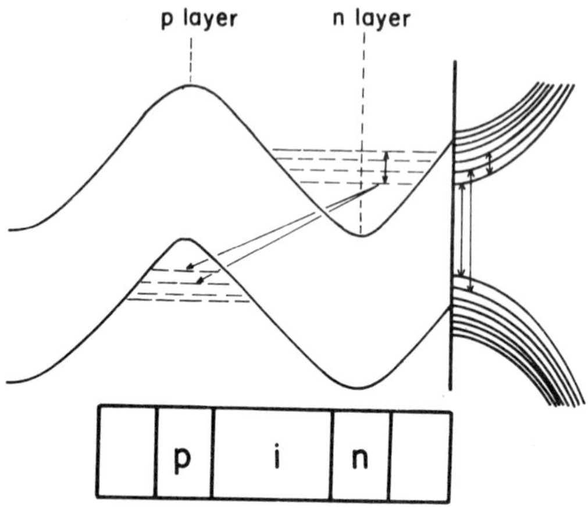

FIGURE 4.24 Band structure and optical transitions in n-i-p-i structure. [After G. H. Döhlev, Doping superlattices (n-i-p-i crystals). IEEE J. of Quantum Electronics QE-22, 1683, 1986.]

Thus, by doping n- and p-type impurities alternately, a superstructure can be obtained that has several exciting properties that might not be obtained with conventional semiconductors. An abrupt and periodic (short distance) variation in the band structure has several important consequences on electrical, optical, and transport properties of the system [81–84]. Since the crystal momentum is conserved in subband transitions, they are observed in the optical absorption spectrum. Similarly, n-n and p-n interlayer transitions are also allowed [81–84]. Thus optical spectra show several unique features, which are summarized below.

There is interaction between the charge carriers located in different layers. Theoretically, it has been found [81–84] that this interaction leads to variation in the amplitude of the superstructure potential, which changes with the intensity of excitation. An increase in the potential reduces the band gap, and the absorption edge shifts toward the low-energy side. This is known as the red shift. Obviously, the magnitude of the shift increases with the level of excitation as more charge carriers are created in different layers. It has also been found that these structures are capable of holding the radiation-generated charge carriers and the absorption coefficient, α, varies with the applied electric field [81–84].

Theoretical studies show that the absorption coefficient for a given photon energy can be increased by orders of magnitude with an applied electric field. On the other hand, the current perpendicular to the layers does not vary with the applied electric field, as the charge carriers cannot move without tunneling and hopping mechanisms. This means that in this direction dark current is expected to be very low. High absorption coefficient (and also one proportional to the applied field) and low dark current is a suitable combination for photodetecting devices.

Band structural features of the n-i-p-i superstructure provide several outstanding properties with respect to the photoconduction (both charge carrier generation and transport) process [83].

In this design, recombination lifetime can be controlled by adjusting spacing and the concentration of the doping. The magnitude of photoresponse increases with the lifetime. It also varies with the applied voltage by two ways. First, applied voltage changes the absorption coefficient and hence the number of charge carriers. Second, the field effect, as in any other detector, increases the photosignal.

Spectral response is also extended toward the low-energy side compared to the band gap of the host material. This is because the intralayer transitions originate from the energy states located below the band gap.

In short, the n-i-p-i structure has several advantages over conventional materials, and even over the compositional superlattice, where the interface defects and lattice matching problems have to be overcome. By adjusting the doping concentration and the region of doping, several basic properties such as carrier concentration, band gap, two-dimensional subband structure, and lifetime of the carriers can be controlled. To understand these strange properties, just a theoretical study is not enough; further extensive experimental investigation in this direction is needed, and only then can the n-i-p-i superstructure be incorporated into modern photodetectors.

4.6. SUPERLATTICE OF DILUTED MAGNETIC SEMICONDUCTORS

Very recently, a combination of magnetic semiconductors and superlattice and multiple quantum well structure was successfully achieved. Notable examples are the $Cd_xMn_{1-x}Te/CdTe$ [96] and $Zn_xMn_{1-x}Se/ZnSe$ [97] systems. In addition to the above-mentioned advantages of the superlattice structure, the system has an additional advantage—the presence of manganese atoms and their localized 3d levels. Their concentration not only helps to control the band gap but also creates additional energy levels originating from the 3d configuration. The position of these levels in the quantum well structure will have a remarkable consequence [97,98].

The diluted magnetic semiconductor superlattice—unlike other superlattices—exhibits [99] a strong exchange interaction similar to that of the bulk materials. Because of this property, spin splitting in DMS layers could be about two orders of magnitude higher than in a non-DMS layer situated approximately 100 Å away. The magnetic splitting could be of the order of the ionization energy of the impurity atom. By adjusting the parameters of the quantum well, it is possible to obtain the energy state of the donor atom just above the energy state in the quantum well. This permits the transfer of electrons from the quantum well to a conducting two-dimensional system. But the situation changes considerably when the magnetic field is applied. The energy state of the donor atom is shifted downward and could reach a position below the energy state located in a quantum well. Electrons from the quantum well will transfer to this state, from where it will not be possible to take part in the conduction process. Thus, a magnetic field induced reduction in conductivity can be expected (yet it has not been proved experimentally). This will have an important consequence on photodetection technology, because dark current is reduced in the magnetic field.

Moreover, the relative positions of the excited states of 3d electrons of the magnetic ion also have significant consequences on the photoconductivity spectrum, as d-d transitions play a crucial role in these materials. Considering the rapid progress of superlattice structures, particularly DMS systems, I believe that it will not be long before the band structural details of these novel materials are understood and they are applied in photonic devices (light emitters and light detectors).

REFERENCES

1. J. Callaway, Quantum Theory of Solids. Academic, New York, 1974.
2. T. S. Moss, Optical Properties of Solids. Butterworth, London, 1959.
3. R. A. Smith, Semiconductors. Cambridge Univ. Press, U.K., 1978.
4. J. I. Pankov, Optical Processes in Solids. Princeton Univ. Press, Princeton, N.J., 1971.
5. D. L. Greenaway and G. Harbeke, Optical Properties and Band Structure of Semiconductors. Pergamon, Oxford, 1968.
6. J. L. Shay and J. H. Wernick, Ternary Chalcopyrite Semiconductors: Growth, Electronic Properties and Applications. Pergamon, New York, 1975.
7. N. Bottka, J. Stankiwicz, and W. Giriat, J. Appl. Phys. 52, 4189 (1981).
8. J. Bardeen, F. J. Blatt, and L. H. Hall, Proc. of Atlantic City Photoconductivity Conference. Wiley, New York, p. 146, 1964.

9. D. G. Thomas, M. Gershenzon, and F. A. Trumbore, Phys. Rev. 133,A269 (1964).
10. D. L. Dexter and R. S. Knox, Excitons (Interscience Tracts of Physics and Astronomy, Vol. 25). Wiley-Interscience, New York, 1981.
11. D. C. Reynolds and T. C. Collins, Excitons. Academic, New York, 1981.
12. R. S. Knox, "Theory of Excitons," Solid State Phys. Suppl. 5. Eds. F. Seitz and D. Turnbull Sect. 9–11, p. 1. Academic Press, New York, 1963.
13. D. C. Reynolds, C. W. Litton, and T. C. Collins, Phys. Rev. 156,881.
14. W. Y. Liang and A. D. Yoffe, Phys. Rev. Lett. 20,847 (1968).
15. Y. Toyozawa and J. C. Hermanson, Phys. Rev. Lett. 21,1637 (1968).
16. H. J. Schulz and M. Thiede, Phys. Rev. B 35,18 (1987).
17. B. N. Figgis, Introduction to Ligand Fields. Wiley, New York, 1966.
18. N. V. Joshi, L. Roa, and A. B. Vincent, Nuevo Cimento 2D, 1880 (1983).
19. M. Gershenzon, D. G. Thomas, and R. E. Dietz, Proc. Int. Conf. on Semiconductor Physics, Exeter Institute of Physics and Physical Society, London, p. 752, 1962.
20. F. Williams, Phys. Stat. Solidi. 25,493 (1968).
21. G. H. Wannier, Phys. Rev. 52,191 (1937).
22. A. S. Davydov, J. Exp. Theoret. Phys. (USSR) 18,210 (1948).
23. D. L. Dexter, Phys. Rev. 84,809 (1951).
24. S. Nikitine, "Excitons" in Optical Properties of Solids. S. Nudelman and S. S. Mitra, Eds., Plenum, New York, p. 197, 1966.
25. R. S. Knox, Theory of Excitons. Academic, New York, 1963.
26. R. J. Elliott and A. F. Gibson, An Introduction to Solid State Physics. Macmillan, London, 1974.
27. P. W. Baumeister, Phys. Rev. 121,359 (1961).
28. K. W. Boer and H. Gutzjahr, Z. Phys. 152,203 (1958).
29. V. V. Ermenko, Sov. Phys. Solid State 2,2315 (1961).
30. M. Trlifaj, Czech. J. Phys. 9,446 (1959).
31. B. V. Novikov, N. S. Sokolov, and S. V. Gastev, Phys. Stat. Solidi(b) 74,81 (1976).
32. A. Corot, J. Ringeissen, and S. Nikitine, in Localized Excitation in Solids. R. F. Wallis, Ed. Plenum, New York, p. 306, 1968.
33. Y. S. Park and D. C. Reynolds, Phys. Rev. 132,2450 (1963).
34. E. F. Gross and B. V. Novikov, in Photoconductivity, H. Levinstein, Ed., Pergamon, New York, 1962, p. 81.
35. E. Gutsche, J. Voigt, and E. Ost, in Proc. of the Third Inter-

national Conference on Photoconductivity. E. M. Pell, Ed. Pergamon, New York, 1969, p. 105.
36. Y. Marfaing, in Handbook on Semiconductors. M. Balkanski, Ed., North-Holland, Amsterdam, 1980, p. 469.
37. R. Planel, C. Benoit a la Guillaume, and A. Bonnot, Phys. Stat. Solidi (b) 58,251 (1973).
38. J. C. Ayache and Y. Marfaing, Phys. Stat. Solidi(a) 2,61 (1970).
39. R. Legros and Y. Marfaing, Phys. Stat. Solidi (a) 19,635 (1973).
40. Y. Toyozawa, Proc. Photoconductivity Conference, Stanford, 1971, p. 151.
41. Y. Toyozawa, in Proceedings of the Third International Conference on Photoconductivity. E. M. Pell, Ed. Pergamon, New York, 1971, p. 151.
42. R. Z. Bachrach and F. C. Brown, Phys. Rev. Lett. 21,685 (1968).
43. J. Dillinger, C. Konak, V. Prosser, and M. Zvara, Phys. Stat. Solidi 29,707 (1968).
44. S. Kurita and K. K. Kobayashi, Tech. Rep. ISSP A430,34 (1970).
45. A. L. Malik, V. I. Livonnov, and Z. D. Kovalyuk, Solid State Commun. 33,621 (1980).
46. N. V. Joshi and B. Vincent, Solid State Commun. 44,439 (1982).
47. N. V. Joshi, L. Martinez, and R. Echeverria, J. Phys. Chem. Solids 42,281 (1981).
48. B. Tell, E. M. Hammonds, P. M. Bridenbaugh, and H. M. Kasper, J. Appl. Phys. 46,2998 (1975).
49. J. M. Martin, N. V. Joshi, and A. B. Vincent, Proc. SPIE 395,267 (1983).
50. C. Kittle, Quantum Theory of Solids. Wiley, New York, 1963.
51. M. Tinkham, Group Theory and Quantum Mechanics. McGraw-Hill, New York, 1964.
52. A. P. Cracknell, Group Theory in Solid State Physics. Taylor & Francis, London, 1975.
53. D. Long, Energy Bands in Semiconductors. Interscience, New York, 1968.
54. E. O. Kane, J. Phys. Chem. Solids 1,249 (1957).
55. H. A. Bethe, Ann. Phys. 3,133 (1929).
56. J. E. Jaffe and A. Zunger, Phys. Rev. B 28,5822 (1983).
57. J. E. Jaffe and A. Zunger, Phys. Rev. B 29,1882 (1984).
58. B. Z. Orlowski, K. Kopalko, and W. Chab, Solid State Commun. 50,749 (1984).
59. W. Kohm and L. J. Sham, Phys. Rev. 140,A1130 (1965).

60. J. L. Shay, B. Tell, H. M. Kasper, and L. M. Schiavone, Phys. Rev. B 7,4485 (1973).
61. N. V. Joshi and R. Echeverria, Solid State Commun. 47,251 (1983).
62. A. Many, Semiconductor Surfaces. North-Holland, Amsterdam, 1971.
63. R. Echeverria, A. B. Vincent, and N. V. Joshi, Solid State Commun. 52,901 (1984).
64. D. Redfield and J. P. Wittke, Proceedings of Third International Conference on Photoconductivity. E. M. Pell, Ed. Pergamon, New York, 1971, p. 29.
65. A. V. Komarov, S. M. Ryabchenko, O. V. Terlestki, I. I. Zheru and R. D. Ivanchuk, Zh. Eskp. Teor. Fiz., 73,608 (1977) translated in Sovi. Phys. JETP, 46,318.
66. J. K. Furdyna and J. Kossut, Eds. "Diluted Magnetic Semiconductors" Vol. 25 of Semiconductors and Semimetals, R. K. Willardson and A. C. Beerg, Academic Press (1988).
67. B. R. Pamplin, N. V. Joshi, and C. Schwab (Eds.), Proc. of Sixth International Conf. on Ternary and Multinary Compounds. Progress in Crystal Growth and Characterization, 10, 1984. Contains several articles on DMS.
68. L. E. Orgel, An Introduction to Transition Metal Chemistry. Butler & Tanner, London, 1966.
69. P. K. Larsen and S. Wittekoek, Phys. Rev. Lett. 29,1597 (1972).
70. Y. Kanai, J. Phys. Soc. Jpn. 24,956 (1968).
71. K. Sato and T. Teranishi, J. Phys. Soc. Jpn. 32,1159 (1972).
72. K. Sato, Y. Yokoyama, and T. Tsushsima, J. Phys. Soc. Jpn. 42,559 (1977).
73. L. Esaki and R. Tsu, IBM J. Res. Develop. 14,61 (1970).
74. L. Esaki, Phys. Colloque C5 45,C5-3 (1984).
75. R. Dingle, W. Wiegmann, and C. H. Henry, Phys. Rev. Lett. 33,827 (1974).
76. R. Dingle, A. C. Gossard, and W. Wiegmann, Phys. Rev. Lett. 34,1327 (1975).
77. E. E. Mendez, G. Bastard, L. L. Chang, L. Esaki, H. Morkoc, and R. R. Fischer, Phys. Rev. B 26,7101 (1982).
78. L. L. Chang, H. Sakaki, C. A. Chang, and L. Esaki, Phys. Rev. Lett. 38,1489 (1980).
79. P. Manuel, G. A. Sai-Halasz, L. L. Chang, C. A. Chang, and L. Esaki, Phys. Rev. Lett. 25,1701 (1976).
80. K. Klitzing, G. Dorada, and M. Pepper, Phys. Rev. Lett. 45, 494 (1980).
81. G. H. Dohler, Phys. Stat. Solidi b 52,79,533 (1972).
82. G. H. Dohler, Phys. Stat. Solidi b 52,533 (1972).
83. G. H. Dohler, IEEE J. Quantum Electron. QE22, 1682 (1986).

84. G. H. Dohler, H. Kunzel, and K. Ploog, Phys. Rev. B 25, 2616 (1982).
85. R. Tsu, L. L. Chang, G. A. Sai-Halasz, and L. Esaki, Phys. Rev. Lett. 34,1509 (1975).
86. F. Capasso, K. Mohammed, and A. Y. Cho, IEEE J. Quantum Electron. QE22,1853 (1986).
87. F. Capasso, K. Mohammed, A. Y. Cho, and A. L. Hutchinson, Phys. Rev. Lett. 47,420 (1986).
88. F. Capasso, Phys. Today 1,S50 (Jan. 1987).
89. L. Esaki and R. Tsu, IBM J. Res. Develop. 14:61 (1970).
90. H. L. Strömer, J. Phys. Soc. Jpn., Suppl. A49,1010 (1980).
91. W. I. Wang, C. E. C. Wood, and L. F. Eastman, Electron Lett. 17,36 (1981).
92. T. Mimura, S. Hiyamizu, T. Fujii, and K. Nanbu, Jpn. J. Appl. Phys. 19,L225 (1980).
93. L. Stormer, A. C. Gossard, W. Wiegmann, and K. Baldwin, Appl. Phys. Lett. 39,912 (1981).
94. H. L. Störmer, A. C. Gossard, and W. Wiegmann, Appl. Phys. Lett. 39,493 (1981).
95. R. Soorykumar, IEEE J. Quantum Electron. QE22,1645 (1986).
96. A. V. Numikko, Y. Hefetz, S. K. Chang, L. A. Kolodziejski, and R. L. Gunshor, J. Vac. Sci. Technol. B4,1033 (1986).
97. L. A. Kolodziejski, R. L. Gunshor, N. Otsuka, B. P. Gu, Y. Hefetz, and A. V. Nurmikko, Appl. Phys. Lett. 48,26 (1986).
98. J. K. Furdyna, J. Vac. Sci. Technol. A4,2002 (1986).
99. R. L. Gunshor, N. Otsuka, M. Yamanishi, L. A. Kolodziejski, T. C. Bonsett, R. B. Bylsma, S. Datta, W. M. Becker, and J. K. Furdyna, J. Crystal Growth 72,294 (1985).

5 Noise

5.1. SIGNIFICANCE OF THE TERM "NOISE"

Noise in the present context is random and unpredictable fluctuations in voltage or in current. This is an unwanted property in any measurement system and normally determines its lower limit. Sometimes the magnitude of random fluctuations is higher than the signal to be handled, and in the case of device applications based on photodetector technology, this leads to incorrect information. If the fluctuations are of the same order as the signal, then, the signal is masked out and/or the desired information is perturbed. The noise consideration is of fundamental importance, particularly when the photoconductivity signal is very weak. An experimental physicist knows that after eliminating the random voltages originating from an improper experimental setup, there is always still noise that determines the measurement capability. In semiconductors, there are several reasons for random fluctuations in voltages.

There are excellent textbooks and review articles [1–10] on noise, and I do not intend to discuss all the aspects of noise and the methods of reducing its effects. We will focus our attention to those types of noise that are very specific to photoconductors and radiation-detecting devices.

Noise in photodetectors degrades the signal, distorts the system performance, and limits the system's capability. Understanding the origin of noise, its magnitude, frequency spectrum, and characteristics and its interrelation with other types of noise is a crucial aspect not only for the design of the photosensor but also for the experimentalist who works in radiation detection, particularly at low levels. The literature shows that noise analysis has helped to improve detector performance by aiding in the selection of a proper chopping frequency, an optimum bias voltage, adequate circuits or filters, and so on. The purpose of this chapter, therefore, is to introduce the types of noise that are frequently observed in photodetectors (both photoconducting and photovoltaic) and to describe them. This chapter will help to explain the importance of noise analysis in improving the limit of detectivity in visible and infrared photodetectors.

5.2. PHYSICAL AND MATHEMATICAL BACKGROUND

Noise is essentially related to the discrete nature of electronic charge. The microscopic behavior of current or voltage is explained with the help of mean densities of free charges (electrons and holes), their thermal velocities, and the average lifetime of charge carriers. There are inherent fluctuations in these parameters as a function of time. Obviously, current or voltage varies as a function of time. A typical example is shown in Fig. 5.1. It is clear that a fundamental study of noise in solid-state devices requires a great deal of the basic formalism of statistical mechanics, a strong knowledge of charge carrier statistics in semiconductors in general, and precise information about transport and recombination processes in a given device.

It can be seen from Fig. 5.1 that the instantaneous value of current I(t) always fluctuates randomly about the mean value of the current. This suggests that the waveform of the noise will not be repeated and cannot be predicted. The variation in the instantaneous value of the current is given by

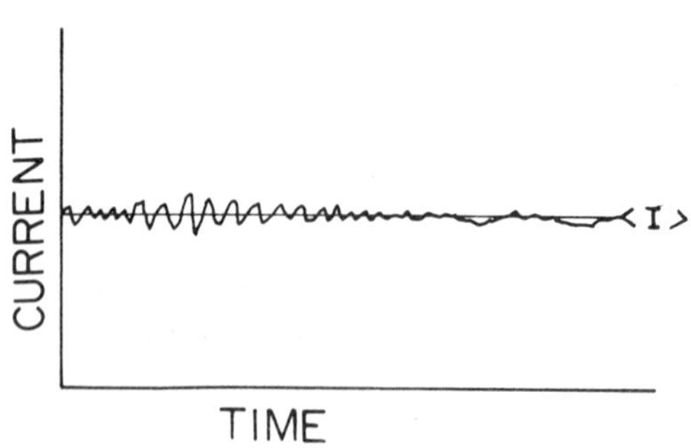

FIGURE 5.1 Current fluctuation I(t) and mean value of current ⟨I⟩.

$$\Delta I(t) = I(t) - \bar{I} \tag{5.1}$$

Since the mean value of ΔI is zero, noise current is defined as the mean square value ΔI and expressed by

$$\langle \Delta I^2 \rangle = \lim_{T \to \infty} \int_0^T \Delta I^2 \, dt \tag{5.2}$$

These fluctuations occur at frequencies from zero to the full spectral bandwidth, and therefore it is necessary to examine noise behavior in the frequency domain rather than in the time domain. Depending upon the type of origin, noise has a typical behavior in the frequency domain. This information is given by the frequency spectrum of noise and is valuable for weak signal detection. This aspect will be discussed later, however, it is worth pointing out that the knowledge of noise mechanisms, their mathematical formalism, and spectral information is not always enough for device engineers; experimental study of noise for a given material (with contacts) and for a specific structure is also needed.

Total noise can be written as

$$\langle \Delta I^2 \rangle = \int_0^\infty S(f) \, df \tag{5.3}$$

where S(f) represents the contribution of current noise at frequency f in unit bandwidth. The same is also true for voltage fluctuations. S is called the spectral density of noise, and if it is independent of frequency, then it is called white noise; a typical example of this noise is Johnson noise.

Photoconductivity originates with the photogenerated free charge carriers that are created by the interaction of photons with matter and annihilated through recombination processes. There are several radiative and nonradiative processes of recombination; they are essentially probabilistic and hence generate noise. The same is true for the generation of photons. In addition to these, there are other sources, such as metal–semiconductor contacts, that also contribute to the noise process.

The presence of traps is reflected in noise in several complicated ways. First, there are fluctuations in the lifetime constant of the charge carriers in the conduction and valence bands. The situation gets more complicated when there are different time constants for different groups of carriers. Second, the mechanisms of capture and ejection of charge carriers and of the interplay between capturing a charge particle from one center and ejecting it into the conduction band or other similar processes are statistical and hence contribute to noise. This aspect will be discussed later.

Independent of the nature of noise, it can be represented in an equivalent circuit as shown in Fig. 5.2. The effects of noise in the circuits are explained by introducing a current or voltage generator. Let $\langle V_n^2 \rangle$ represent the mean square value of noise in a bandwidth df; then the spectral response is given by

$$\langle V_n^2 \rangle = S(f)\ df \tag{5.4}$$

In certain cases, Eq. (5.4) gives a proper method for handling noise in photodetecting devices.

Noise is classified according to its origin, and we will consider here only the types of noise that frequently occur in photoconducting devices.

5.3. TYPES OF NOISE

5.3.1. Photon Noise

Photon noise is a fundamental type of noise that originates from the random generation of charge carriers. It is a limiting noise in detectors and gives the theoretical limit for NEP (noise equivalent power) and detectivity. The physical principles involved in this type of noise are rather simple.

If radiation of power P is incident on the photoconductor, then the average rate $\langle r \rangle$ of production of charge carriers is given by (see, for example, Ref. 10)

$$\langle r \rangle = \frac{\eta P}{h\nu} \tag{5.5}$$

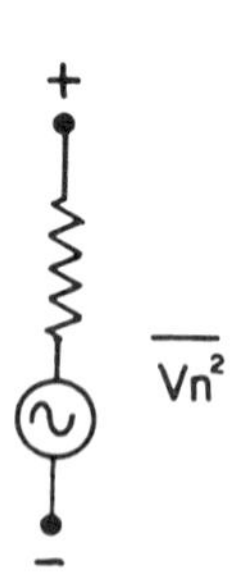

FIGURE 5.2 Noise representation associated with current fluctuation.

The significance of the symbols has been described already. $P/h\nu$ represents the number of photons per unit time incident on the detector. The events of the charge formation are randomly distributed in time, and the mechanism of formation is a probabilistic process. For a large number of observational phenomena such as this, the generation process is expressed by a Poisson distribution, according to which the probability of creation of N electrons in a time interval τ is given by [11–13]

$$P(N\tau) = \frac{(\langle r\rangle\tau)^N}{N!} e^{-\langle r\rangle\tau} \tag{5.6}$$

We know that for a Poisson distribution, the mean square fluctuation of the number of events in a fixed interval of time is

$$\langle [N - \langle N\rangle]^2\rangle = \langle N^2\rangle - \langle 2N\langle N\rangle\rangle + \langle N\rangle^2 \tag{5.7}$$

Using a property of Poisson distributions,

$$\langle N^2\rangle - \langle N\rangle^2 = \langle N\rangle \tag{5.8}$$

we get

$$\langle (N - \langle N\rangle)^2\rangle = \langle N\rangle \tag{5.9}$$

This is a very useful property, which a Poisson random variable possesses. It states that the expectation value is equal to the variance (for proof see Ref. 11,12).

The fluctuations in noise occur at frequencies from zero to full spectral bandwidth. The noise spectrum can be given as

$$\langle I_N^2(f)\rangle = 2\langle r\rangle |i(\omega)|^2 \tag{5.10}$$

where $I_N(f)$ is the spectral density of the mean square noise current and $i(\omega)$ is the Fourier transform of the current pulse given by [9]

$$i(\omega) = \int_{-\infty}^{\infty} I(t)e^{-i\omega t}\, dt \tag{5.11}$$

Here $I(t)$ is the current as a function of time.

Equation (5.9) shows how the fluctuations in the charge carriers are created by the flux of photons. (This is also true when the photon flux varies with time.)

It is also clear from Eq. (5.9) that photon noise originates from a statistical variation in the number of photons arriving in a fixed time interval. We know that mean photon deviation is equal to the square root of mean photon number in a short interval of time. Thus, the number of photons and hence the number of charge carriers created in a given photoconductor vary from one time interval to another. This creates fluctuations at the output of a photodetector, which are classified as photon noise. The mean square fluctuation of the incident power per unit bandwidth is given by [10]

$$\langle I_n^2 \rangle = (I - \langle I \rangle)^2 = \frac{e^2 \langle n \rangle}{\tau^2} \tag{5.12}$$

The mean square noise current measured in a time interval τ is

$$\langle I_n^2 \rangle = \frac{e^2}{\tau^2}\left(\frac{I_n \tau}{e}\right) \tag{5.13a}$$

and the average noise current can be written as

$$\langle I_n^2 \rangle = eI/\tau \tag{5.13b}$$

The mean square current fluctuation, averaged over many independent measurements (this is a very realistic situation) is expressed by

$$\langle I_n^2 \rangle = 2eI_n \; \Delta f \tag{5.14}$$

where $1/2\tau = \Delta f$. The spectral density in this case is equal to $2eI_n$ [10]. This is the well-known shot noise first observed in vacuum diodes and now frequently found in several types of photodiodes. As mentioned in Chapter 1, the S/N ratio and NEP are two important parameters in photodetectors. With the help of the basic equations, (5.13) and (5.14), we can evaluate these parameters easily for an ideal photodetector.

Let us consider a common laboratory situation where the detector is looking at a narrow spectral range, that is, background radiation is negligible. In this case, the signal current I_s generated due to radiation of power P is given by

$$I_s = \frac{\eta P}{h\nu} e \tag{5.15}$$

The mean square noise, using Eq. (5.14) for I_S is

$$\langle I_n^2 \rangle = 2e^2 \frac{\eta P}{h\nu} \Delta f \tag{5.16}$$

The ratio of the signal power to noise power is

$$S/N = \frac{\langle I_s \rangle^2}{\langle I_n \rangle^2} = \frac{\eta P}{2h\nu\Delta f} \tag{5.17}$$

Therefore, the noise equivalent power, NEP, becomes

$$NEP = 2 \frac{h\nu}{\eta} \Delta f \tag{5.18a}$$

since $\Delta f = 1/2\tau$, for unit quantum efficiency, NEP is

$$NEP = \frac{h\nu}{\tau} \tag{5.18b}$$

Thus, in principle, it is possible to detect on the average one single photon per unit measurement interval. Here we have assumed that the contribution from other types of noise is negligible, and in such circumstances detectivity is limited by photon noise. This is certainly an ideal situation.

5.3.2. Johnson Noise

Johnson noise is associated with the fluctuations of the thermal velocities of free charge carriers and is therefore sometimes called thermal noise. However, "Johnson noise" is a more suitable term as there are other types of noise whose origin is directly or indirectly related to thermal fluctuations. Johnson noise appears at the terminal of every resistive material; naturally it appears at the output of the photodetector. Its physical origin can be understood very easily. In a photodetector or in any resistive material, electrons or holes are in a state of continuous agitation; their energies are determined by Fermi–Dirac statistics. Each electron (hole) travels from one electrode to the other by a series of random paths. At the end of each small trajectory, a collision takes place with a phonon or an impurity atom, and the energy and direction of the electron (hole) after the collision are entirely different than before. As a result of the combined effects of all the electron transit parameters, there exist local fluctuations in the space charge. As a consequence of this, the open circuit noise voltage is created at the end of the sample. The noise originating from this process obviously depends on the thermal velocity of charge carriers, that is, it is proportional to the absolute temperature T. Johnson noise voltage per unit bandwidth across any resistor R is given by

$$\langle V_n^2 \rangle = 4kRT \tag{5.19a}$$

where k is the Boltzmann constant. Noise power in this case is given by

$$P_n = kT\ \Delta f \tag{5.19b}$$

It is worth pointing out that equation (5.19) has been obtained from classical theory and is valid up to 10^{13} Hz. For low temperatures and high frequencies, Eq. (5.19b) is not valid. By using a quantum-mechanical treatment it has been found that Eq. (5.19b) is modified and becomes [14]

$$P_n(f) = \frac{h\nu\ \Delta f}{e^{h\nu/kT} - 1} \tag{5.19c}$$

It is clear that except at very high frequencies or at very high temperature, Eq. (5.19c) converts into the conventional Eq. (5.19b). Since high-speed photodetectors have become available, noise studies in such detectors demand the use of Eq. (5.19c).

It is worth mentioning that in devices like n-p junctions, the charge carriers in different regions have different densities and different velocities. Obviously, their contributions to Johnson noise should be estimated separately.

During experiments, it is frequently observed that Johnson noise originates from two different resistors maintained at different temperatures. A typical example is a photodetector of resistance R operated at T (e.g., liquid nitrogen or liquid helium temperature). Effective noise voltage can be calculated by assuming that each resistor in a network acts as a noise source of magnitude given by Eq. (5.19). The mean square value of the total fluctuation voltage can be calculated for series and parallel resistances with the help of Fig. 5.3.

When the resistances are in series, the effective noise voltage can be obtained directly and is given by

$$V_n^2 = 4k\ \Delta f\ (R_1T_1 + R_2T_2) \tag{5.20}$$

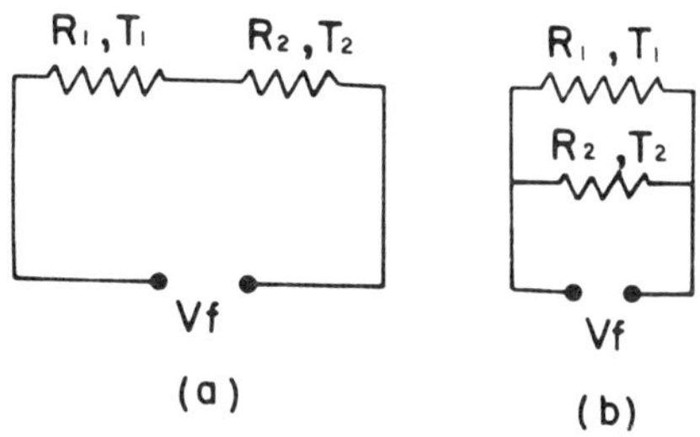

FIGURE 5.3 Resistances R_1 and R_2 (a) in series and (b) in parallel operated at temperatures T_1 and T_2.

When they are in parallel, the noise can be obtained by considering the noise current due to R_1 and R_2 resistances. The total noise current is given by

$$I^2_{total} = I_1^{\ 2} + I_2^{\ 2} \qquad (5.21)$$

since $R_T^{-1} = R_1^{-1} + R_2^{-1}$.

The magnitude of the noise voltage is

$$V_n^{\ 2} = 4k\Delta f \frac{R_1 R_2 (R_1 T_2 + R_2 T_1)}{(R_1 + R_2)^2} \qquad (5.22)$$

This is a very useful equation for estimating Johnson noise, particularly for infrared detectors, because they are generally operated at low temperature with the load resistance at room temperature.

5.3.3. Generation-Recombination Noise

Generation-recombination (g-r) noise is originated due to the fluctuations in the number of free charge carriers generated randomly through traps. The processes of capture and reejection of charge carriers are accelerated with increase in temperature and with the absorption of photons. Generation-recombination noise is present not only in photosensitive devices but also in nonphotosensitive devices, but it is predominant in the former and limits the detectivity of several commercially available detectors. Separate studies of some of these detectors have been carried out [9,15,16]. One of the important properties for determining detectivity as a function of frequency is the power spectrum of g-r noise, which can be estimated as follows.

Let us consider, for simplicity, a piece of n-type semiconductor of length l, cross section A, and mean free electron density n.

The conductivity of the rod is $\sigma = n_0 e \mu_n$, and the resistance R is given by

$$R = 1/ne\mu_n A = l^2/N_0 e\mu_n \tag{5.23}$$

where N_0 is the total number of electrons within the sample and is equal to Aln_0. The current passing through the sample when the voltage V is applied is

$$I = V/R = \frac{V_0 N_0 e \mu_e}{l^2} \tag{5.24}$$

since the mobility, μ_n, is constant, the fluctuation in the current is caused only by the variation in N_0.

$$\Delta I = \frac{\Delta N}{N_0} I \tag{5.25}$$

ΔN is the random change originating from g-r processes. ΔI is the noise current, and its mean square value is given by

$$\langle \Delta I^2 \rangle = \frac{\langle \Delta N^2 \rangle}{N_0} I^2 \tag{5.26}$$

It can be shown that the variation in the electron concentration is [17]

$$\langle \Delta N^2 \rangle = g_0 \tau_n \tag{5.27}$$

where g_0 is the generation rate of the electrons and τ_n is the mean lifetime of the electrons in the conduction band. In this case, the frequency spectrum is given by [5,17] (for pedagogical discussion see Ref. 5)

$$S(f) = 4I^2 \frac{\langle \Delta N^2 \rangle}{N_0^2} \frac{\tau_n}{[1 + (2\pi f \tau_n)^2]} \tag{5.28}$$

According to Eq. (5.28), the noise spectrum is flat for low frequency, $\omega\tau_n \ll 1$, and falls inversely at high frequency, $\omega\tau_n \gg 1$.

As mentioned earlier, g-r noise is present in dark as well as in illuminated semiconductors. In the former it is caused by thermal generation of charge carriers, and for a photoconductor under illumination the major contribution comes from the number and the lifetime of electron—hole pairs created by the absorption of photons. Assuming that the recombination processes also follow the Poisson distribution (which is not always true), the noise spectrum can be given as [18,19]

$$\langle I_n^2 \rangle = 4e \frac{I_{pho}}{M(0)} \left(\frac{\tau_V}{\tau_{transit}} \right)^2 \frac{\Delta f}{1 + \omega^2 \tau_V^2} \tag{5.29}$$

where $\tau_{transit}$ is the transit time of the electron, τ_V is the pair lifetime, M(0) is the gain of the device, and I_{pho} is the steady-state value of the photocurrent. Equation (5.29) relates the noise to gain at zero frequency, transit time, and apir lifetime τ_V. But dc gain is related to the dynamic gain M(w) by the following equation [18]:

$$M(w) = \frac{\tau_n}{t_{transit}} (1 + \omega^2 \tau_n^2)^{-1/2} \tag{5.30}$$

substituting Eq. (5.30) in (5.29), the spectral distribution is

$$\langle I_n^2 \rangle = 4eI_{pho} \frac{M^2(\omega)}{M^2(0)} \Delta f \tag{5.31}$$

Equation (5.31) suggests that by measuring photocurrent and noise spectrum it is possible to obtain the dynamical gain M(w) of the device.

Vilcot et al. [19] carried out such an investigation on a GaAs planar photoconductive detector. The noise spectra were obtained in the dark and under illumination, and g-r noise due to illumination was estimated by subtracting the dark contribution from the noise under illumination. Using Eq. (5.31) the dynamical gain was obtained. This value was confirmed by other independent methods such as the picosecond response technique. This work is being cited here to emphasize that noise analysis not only helps to explain the capability of the detectivity but can also provide valuable information about the key parameters (such as gain as a function of frequency) of modern photodetectors.

As mentioned earlier, g-r noise is predominant and very often determines the limit of detectivity. A typical example is the infrared photoconducting detector $Hg_xCd_{1-x}Te$ [15,16]. Theoretical and experimental investigations have been carried out by Smith [20], who considered the effects of the drift and diffusion of minority carriers on the charge-neutral fluctuations in the electron/hole density. These fluctuations are reflected in the noise voltages.

A detailed study reveals that g-r noise in this detector depends on the biasing voltage also. For low applied voltages (up to 20 V/cm), g-r noise increases almost linearly with voltage, but after this value its magnitude remains constant. The frequency study shows that for GaAs the noise is nearly constant up to 10^2 Hz and then drops very quickly as the frequency increases [5,20]. A separate study for other materials and devices is required. These factors are important for determining the specific detectivity and detection of weak signals with $Hg_{1-x}Cd_xTe$. This is the predominant type of noise in photoconducting devices, and specific information for a particular case must be obtained separately. Generation-recombination noise obviously depends upon the trap and recombination processes, which vary considerably from material to material and from device to device. Therefore, special consideration is given to these aspects shortly.

5.3.4. Modulation Noise

The term "modulation" refers to the fluctuations in charge carrier density (and hence conductivity) other than thermal fluctuations in the statistical distribution of carriers [17]. The exact mechanism of this type of noise is not clear yet, but it originates from the contribution of surface defects or flaws in the modulation of current. In principle, surface defects can be reduced or eliminated, and therefore this is not a fundamental type of noise. The noise power spectrum is given approximately by [21]

$$S_i(f) = \frac{C_1 I^2}{f^a} \qquad (5.32)$$

where C_1 is a constant whose value can be determined experimentally and "a" is nearly equal to 1. In a well-prepared semiconductor surface, the magnitude of this noise is smaller than that of the other types of noise and its presence can be neglected.

The frequency range over which modulation noise is observed is extensive. In germanium wire it has been observed for frequency as low as 10^{-4} Hz [22], whereas Hyde [23] observed it in germanium crystal at 10^6 Hz. This shows how strongly the form of the material influences the modulation noise spectrum.

5.3.5. Contact Noise

Semiconductor—metal contact is an important issue and a source of noise in photoconductivity. We know that even ohmic contacts generate noise, and the frequency spectrum of contact noise is given by [14]

$$S(f) = \frac{C_2 I^b}{f^a} \tag{5.33}$$

where the values of a and b are very close to 1 and 2, respectively. C_2 is generally obtained experimentally. The frequency spectrum of this type of noise is similar to the modulation noise spectrum except that the values of the constants are different. The magnitude of the noise can be reduced by changing the contact metals and improving the technique of making them. Very often it is difficult to find an adequate material and the technique for making an ohmic or Schottky contact as the case may be. In this case, the magnitude of contact noise is high enough to distort the photoconductivity measurements.

It is clear from Eq. (5.33) that the contact noise reduces as frequency increases. This suggests that perturbation from contact noise in photocurrent can be reduced by using a phase-sensitive technique with high chopping (modulating) frequency [24].

Comparison of (5.32) and (5.33) shows that noise spectra for both types of noise (contact and modulation) are similar. Obviously, it is difficult to evaluate them separately. However, knowing that contact noise predominantly appears near the contact region whereas moodulation noise appears throughout the entire region, the latter can be separated out by measuring the noise at four points.

5.3.6. 1/f Noise

The precise origin and mechanism for the noise known as 1/f noise, even in metals, is an open question and still remains a central problem in the physics of fluctuations [7]. A full grasp of it is not expected for working with photodetecting devices. Here, we will mention briefly the major processes that contribute to 1/f noise and their role in radiation detection. Such an understanding is necessary to reduce noise levels to improve NEP and detectivity. It is worth mentioning that mathematical expressions for estimating the magnitude of noise are available only for special cases [3,6,7].

This type of noise is generally observed in semiconductor devices, and it was once thought that it was due to bulk properties. Recently, it has been observed that 1/f noise is completely absent from silicon junction field effect transistors. This suggests that it is not a true bulk effect [25]. It originates at the semiconductor interfaces where the surface states play a predominant role in the trapping and detrapping [3] of the charge carriers and fluctuations in the surface recombination velocities are very common [26,27]. 1/f noise is also commonly observed in thin-film resistors.

Surface states, flaws, and traps, under certain conditions, create spontaneous fluctuations in carrier density (by trapping or retrapping the charge carriers). In germanium it has been found that this type of noise is higher when the inversion layer is formed at the surface and lower when the accumulation layer is present [3]. The presence of moisture and the adsorption of certain gases also increases 1/f noise. This is because the fluctuating occupancy of the oxide traps (or the traps created by the adsorbed gas) modulates the free carrier scattering at the surface. This process produces mobility fluctuations of charge carriers [17]. Unfortunately, the pertinent aspects of the physics and chemistry of adsorbed gases are not all completely clear yet, which complicates our understanding of mechanisms for 1/f noise.

1/f noise is commonly observed in p-n junction diodes [28–35] and in photodiodes, particularly those operated in reverse bias. A notable example is the $Hg_xCd_{1-x}Te$ photodiode [34–35]. Experimental investigations confirm that surface, bulk, and photogenerated bulk diffusion currents contribute significantly in 1/f noise. Several models have been proposed to explain 1/f noise quantitatively [28]. Present theories are advanced enough to estimate the magnitude with reasonable accuracy for certain cases such as for photodiodes where the diffusion current or g-r mechanisms are dominant [28]. The last aspect is particularly important because of its direct effect on infrared detection technology.

The magnitude of 1/f noise in a particular device depends upon the growth method and processing technology. In a device where the ion implantation technique is carried out, 1/f noise is expected to be the dominant factor [35]. Experimental study shows that this type of noise varies with temperature, junction area, and applied diode voltage [35].

Very recently a correlation between 1/f noise and grain boundaries in thin gold films has been reported [36]. Experimentally it has been confirmed that the sample with the lowest number of grain boundaries has the lowest 1/f noise. In short, variation in density and modulation together with the interaction of mobile charge carriers with grain boundaries are also factors in 1/f noise. This suggests that a detector fabricated of polycrystalline material can be expected to exhibit this type of noise predominantly.

It is worth pointing out that in devices such as p-n photodiodes, the noise is not always strictly explained by the 1/f form but it is very close to it in some frequency range and the real form is rather complex. A careful study of noise in the InGaAs/InP photodiode [37,38] shows that the magnitude and noise spectra of dark and photocurrent for a particular device depends upon the direction and magnitude of the applied voltage and the wavelength of the incident radiation. In the majority of cases, white noise dominates below some critical voltage, while above that voltage 1/f noise dominates. These results are generally explained on the basis of an accumulation of charge carriers at the interface or a high concentration of defect states close to the interface.

The above discussion shows the importance of surface preparation of a photoconductor, a reliable technique of epitaxial growth, and the construction of defect-free p-n junctions. Considering both the complexity of the fabrication process and the contribution from several types of currents, it is expected that the magnitude of 1/f noise should vary from sample to sample and device to device. Therefore, the experimental investigation for a specific detector is highly desirable for low-level radiation detection.

A demonstrative example of the effects of 1/f noise and white noise on detectivity has been presented by Kleinpenning [28] in photovoltaic (PV) and extrinsic photoconductive detectors. In PV detectors, low-frequency current noise is determined by examining 1/f noise, and in such cases it has been found that to achieve maximum D* the detector temperature should be decreased with increasing wavelength. This and similar observations should not be generalized, as the origin of 1/f noise in photodetectors, even today, is an open issue.

5.3.7. Shot Noise

Shot noise was first discovered in thermionic valves; it originates from the fluctuations in the rate at which the electrons are ejected from the cathode. The variation in the emission rate of electrons is reflected in the current fluctuations. Shot noise appears in semiconductor devices, particularly in n-p junctions and the boundary between the metal and the semiconductor [37,38]. The flow of charge carriers across the potential barrier is not constant as the crossing of individual carriers occurs independently and at random. Therefore, this process gives rise to shot noise. Of course, the separate contributions of holes and electrons should be estimated in these devices. Under the assumption that the time required to cross the barrier is smaller than the lifetime of the charge carrier, the spectral density is given by [17]

$$S(f) = 2eI\left[\frac{\sin \omega\tau/2}{\omega\tau/2}\right]^2 \tag{5.34}$$

When $\omega\tau \ll 1$, the equation reduces to 2eI, which is the well-known low-frequency shot noise, for thermionic valves.

Shot noise plays a predominant role in those devices in which the charge carriers have to pass a potential barrier. In the next chapter, we will see modern multilayer photodiodes where consideration of shot noise is fundamental.

5.3.8. Current Noise

Johnson noise, as mentioned earlier, originates from the thermal fluctuations of charge carriers at equilibrium. When the current is passed through the material (semiconductor or metal), charge carriers are no longer in equilibrium and another type of noise associated with the nonequilibrium condition, different from Johnson or shot noise, appears. The noise voltage is proportional to the magnitude of the current and is prominent at low frequencies. For higher frequencies, shot or Johnson noise predominates over current noise. Figures 5.4 and 5.5 show the variation of rms noise voltage $V^2_{current}$ to $V^2_{Johnson}$ noise for a carbon resistance for various values of current and frequencies [6].

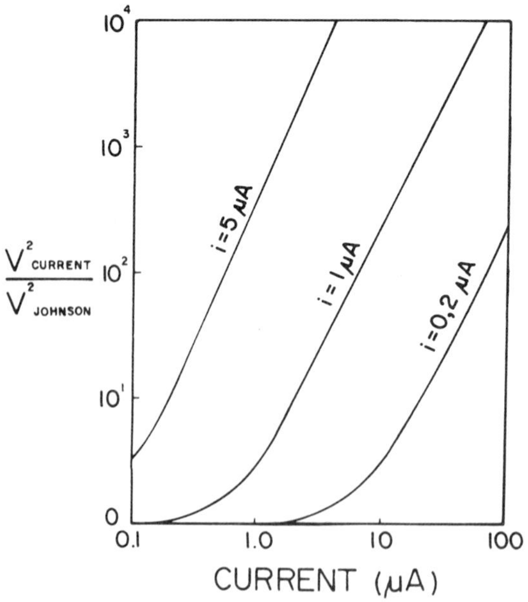

FIGURE 5.4 Ratio of the square of current noise to square of Johnson noise for various values of current. [From Ref. 6. Reprinted by permission.]

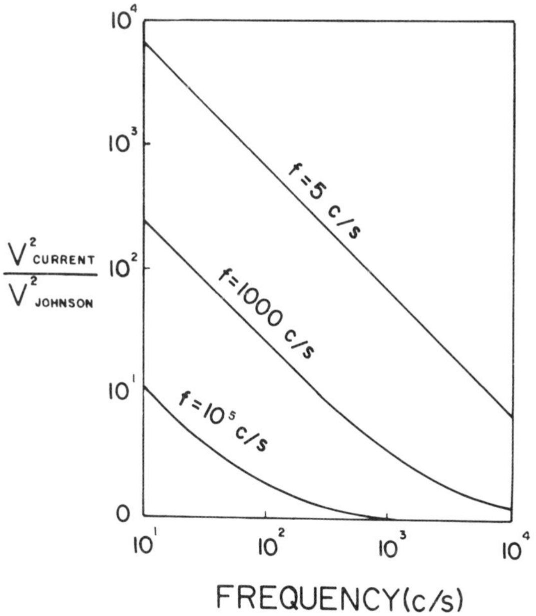

FIGURE 5.5 Ratio of the square of current noise to square of Johnson noise for various frequencies. [From Ref. 6. Reprinted by permission.]

The importance of (dark) current noise was discussed in Chapter 1, and we know that S/N ratio (and therefore detectivity) depends on it. The magnitude of dark current noise and its spectral distribution cannot be estimated theoretically, and a separate experimental investigation must be carried out for each particular device [39–41]. We will shortly see that a careful analysis of dark current noise provides very useful information [42].

There is another interesting aspect of current noise. Unlike many other types of noise, it depends upon the dimensions and the geometry of the material, even though the material is homogeneous. A rough estimation of current noise for a macroscopic sample shows that for a wire of radius r and length l the magnitude of noise is given by [5,6]

$$V_n^2 \propto \frac{I^2 l}{r^b} \tag{5.35a}$$

and for a uniform rectangular layer of length l, breadth b, and thickness d, it is

$$V_n^2 \propto \frac{I^2 l}{b^3 d^3} \tag{5.35b}$$

Equations (5.35a) and (5.35b) show the importance of the form and size of the sample.

I cite this information here because modern photodetecting devices are made with epitaxial thin layers and the dimensions of the devices are as small as a few micrometers (see Chapter 6). Obviously, in this case current noise depends upon the structure and geometry of the device. It is quite possible that in such devices current noise could be the limiting factor rather than g-r noise. The only way to know is by a theoretical and experimental investigation for a specific system. There is a real need for such a study for photosensitive devices integrated monolithically as the dimensions of these devices are of the order of a few micrometers. The same is true for devices based on the superlattice structure (see Chapter 6), where the active thickness is of the order of a few hundred angstroms. Fortunately, some work in this direction has been done, and useful information such as the optimum bias voltage for detectivity measurements has been reported [42—44]. We will refer to this aspect when we will discuss the practical utility of noise analysis in some modern sensors in Chapter 6.

5.3.9. Photocurrent Noise

Photocurrent noise is a combination of the various types of noise mentioned above. A fairly complete discussion of several of its aspects is given by Rose [45]. The formalism used by Rose is rather simple, yet it helps to visualize the physical origin of the process. Of course, the details of the mechanism such as recombination processes, photon statistics, and trap contribution to the noise are avoided in his treatment for simplicity.

The relation between the magnitude of photocurrent and the noise associated with it can be estimated in a simple way. The photocurrent I_{pho} is given by [45]

$$I_{pho} = \frac{e g_0 \tau}{t_{transit}} \tag{5.36}$$

where $t_{transit}$ is the transit time of the electron from cathode to anode. Sometimes the gain, G, of the system is defined as

$$g = \tau / t_{transit} \tag{5.37}$$

As we have seen, the generation of charge carriers is a random process and the noise associated with it is given by Eq. (5.12). Thus, the noise current in terms of g_0 can be written as

$$\langle I_n^2 \rangle = g_0 e^2 / \tau \tag{5.38a}$$

since $1/2\tau = \Delta f$, Eq. (5.38a) becomes

$$\langle I_n^2 \rangle = 2 g_0 e^2 \ \Delta f \tag{5.38b}$$

Here we have considered only the noise from the charge generation process. Electrons from the conduction band are recombined with the holes from the valence band by a random process, and thus the noise is generated. Assuming that it is on the same order as the photogeneration process [45] (in fact it depends on the mechanisms of recombination and the details will be discussed shortly), the noise in the photoconductor due to g-r processes becomes

$$\langle I_n^2 \rangle = 4 g_0 e^2 \ \Delta f \tag{5.39}$$

In a system of gain G, the charge contribution per event is eG, and hence replacing e by eG and using Eqs. (5.36) and (5.37), (5.39) becomes

$$\langle I_n{}^2 \rangle = 4 e I_{pho} G \ \Delta f \tag{5.40}$$

that is, it is equal to the number of electrons passing through the photoconductor per excitation. The transit time and hence noise depend on the applied voltage. Further discussion on this aspect can be found in Rose [45].

In the presence of traps the situation is slightly different. Some of the excited electrons are captured in the traps, which are strongly coupled to the conduction band via thermal excitation; that is, some of the electrons from traps are ejected to the conduction band, where they take part in the conduction process. This means that the time spent by the electrons in the conduction band is not the same as it would be if the traps were not present. The expression for the mean square noise is the same, but the bandwidth Δf is altered.

In short, the effect of traps is to reduce the bandwidth from $1/2\tau_n$ to $1/2\tau_{trap}$, where τ_{trap} is the effective time spent by electrons in the traps. The numerical value of the gain is changed but the gain–bandwidth product remains constant.

According to Eq. (5.40), the mean square noise should be reduced along with the photocurrent. This is not a common observation in those photoconductors where the response is dominated by traps. Even at a low value of photocurrent, considerable fluctuations are observed in it. For lower values of photocurrent, the noise increases, suggesting that another mechanism is dominant at a low value of photocurrent and in the presence of traps. This behavior will be discussed shortly.

The above formalism proposed by Rose [45] is not sufficiently explicit and cannot be useful for evaluating the magnitude of the noise in any photoconductor or in any specific device as it does not take into account the details of the recombination processes and the mechanisms of trapping and emptying defect states. All these are random events that contribute to the photoconductivity noise. The magnitude of noise and its spectrum depend upon the specific mechanism and hence should be discussed separately. Some typical illustrative examples are given below.

One of the most frequently used devices for photodetection is a photoconductive cell. The noise behavior in these devices has been studied extensively [5,6,36–39]. There are three types of intrinsic photocells, and their dominant noise mechanisms are different. Table 5.1 shows the noise mechanisms that dominate specific detectivity in some well-known photosensors. We know that g-r noise is a predominant type, but the specific processes of recombination vary and this makes a great difference in both the magnitude and the spectrum of noise.

First, let us consider a relatively simple case where the contribution of traps is negligible and the photoconductor is working in or near the intrinsic range [46–48]. The photoconductance is carried out by nearly equal numbers of electrons and holes. Typical examples are Si (1.1 μm), Ge (1.65 μm), InAs (3.8 μm), and InSb (7.1 μm). Assuming that the g-r mechanism is dominant and that Schockley-Read centers are present, the noise spectrum in these materials is given by [2]

$$S_n(f) = S_p(f)$$

$$\simeq \frac{4n_0p_0Ad\,\tau_{SR}}{(n_0 + p_0)(1 + \omega^2\tau_{SR}^2)} + \frac{4n_0p_0Ad\,\tau_{p0}(n_0 + n_1)}{(n_0 + p_0)\tau_{SR}\,\tau_{n0}(p_0 + p_1)}$$

$$\times \frac{\tau_2}{1 + \omega^2\tau_2^2} \tag{5.41}$$

where

TABLE 5.1 Dominating Noise Mechanisms in Certain Photoconducting Sensors

Photoconducting	Operating temperature (°K)	Noise mechanism that limits detectivity
PbS	77	Current
PbS	300	Current
PbSe	77	Current
PbSe	300	Current
InSb	77	Current
	300	Thermal
Ge(Au, Zn, or Hg)	≅ 70	Current (for low frequencies)
		g-r (for high frequencies)
CdHgTe	≈ 70	g-r
Si(Zn)	273 and 77	g-r (for low and for high frequencies)

$$\tau_{SR} = \frac{\tau_{n0}(P_0 + \underline{p}) + \tau_{p0}(n_0 + \underline{n})}{n_0 + p_0}$$

and

$$\tau_2 = \frac{\tau_{n0}\,\tau_{p0}\,N_t}{\tau_{SR}(n_0 + p_0)}$$

Here A is the area and d is the active thickness of the detector, which is roughly equal to light penetration depth in the sample. It is worth pointing out that this formula is not valid for superlattice structures where the active region is a few hundred angstroms. This shows that the noise and hence NEP and detectivity (also specific detectivity) are functions of the modulating frequency ω.

In the second type of photodetectors the sensitivity is controlled by the presence of traps, which capture the minority carriers. Typical examples are lead compounds such as PbS, PbTe, and PbSe.

The noise in these detectors has been studied in detail [49,50]. The precise information for a specific detector can also be obtained from the manufacturer's data sheet. In general, the generation-recombination noise spectrum for electrons and for electrons and holes is estimated on the basis of a Markov process, which can be expressed as [51]

$$S_n = \frac{4n_0 \tau_1 V}{1 + \omega^2 \tau_1^2} \cdot \frac{n_0^2}{(p_0 + p_1 + n_c)p_0} + \frac{4n_0 \tau_{2n} V}{1 + \omega^2 \tau_2^2} \tag{5.42}$$

and for electron—hole interaction,

$$S_{n\text{-}p}(f) \simeq \frac{4n_0 \tau_1 V}{1 + \omega^2 \tau_1^2} \cdot \frac{n_0}{p_0 + p_1 + n_c} - \frac{4n_0 \tau_{2n} V}{1 + \omega^2 \tau_2^2} \cdot \frac{c_{p0}}{C_n(N_t - n_c)} \tag{5.43}$$

where V represents the volume of the detector in which a g-r process is operating. τ_{1n} and τ_{2n} are given by

$$\tau_{1n} = (N_t - n_c)/cn_1(p_0 + p_1 + N_t) \tag{5.44a}$$

and

$$\tau_{2n} = [c_n(N_t - n_c)]^{-1} \tag{5.44b}$$

These expressions include the effects of the presence of traps. The density of empty and occupied traps appears explicitly, as expected, in these equations. The parameters of the traps such as trapping rates are included in the definition of τ_{1n} and τ_{2n}.

Equations (5.41) and (5.42) show how the noise spectra depend on the particular mechanism of the recombination, the role of the parameters of the trap, and the lifetime of charge carriers.

The third type of detectors are those in which two types of traps are active, one for the majority carriers and the other for the minority carriers. Examples of these types of detectors are wide-band-gap photodetectors such as CdS, ZnS, and CdSe. It is widely accepted that the photosensitivity in these materials is due to the intentionally introduced impurity in the band gap. The role of a set of impurity levels is to increase the lifetime of the charge carriers and hence the photoresponse [52,53].

In the presence of two optically active traps, the calculation of the noise spectrum is further complicated. I do not intend to give a full account and formal derivations of noise spectra here, as those are available in the literature [2]. Equations (5.41)—(5.43) are just mentioned to emphasize the complications that could arise in evaluating a noise spectrum for a specific type of recombination process. This also shows the limited applicability for the model proposed by Rose [45].

In these materials the lifetime varies from 10^{-6} to 10^{-2} sec and the mathematical formalism of the noise spectrum is controversial [54, 55]. Large fluctuations are sometimes observed that cannot be understood on the basis of a Poisson distribution, and there are some experimental details that are not explained satisfactorily. In these cases, the magnitude of the noise and its spectrum should be determined experimentally.

In extrinsic photodetectors, the noise mechanisms and the calculation of the noise spectrum are more complicated. They depend upon the types of donors or acceptors and their ionization energies. This sensor may contain double acceptors, and in that case the rate equation is more complex. In addition, the Fermi—Dirac statistics is also modified. The role of neutral shallow donors in the noise process is understood, and it has been found that low-frequency ($f < 50$ Hz) photocurrent noise is generated through the breakdown or freeze-out of neutral donors. A typical example is chromium-doped GaAs, where noise of this type is reported at 4.2 K [56]. If noise mechanisms in different materials with different impurities at different temperatures are analyzed, then it can be realized that there is no end to the possibilities. Therefore, a general remark about magnitudes (even relative magnitudes) and the frequency spectrum is out of the question. A rule of thumb is to carry out experimental investigations of the noise spectrum on a particular device rather than to use theoretical knowledge when several unknown parameters are involved.

The approach taken above to explain the noise spectrum of intrinsic photodetectors is also approximate. In fact, there exists the possibility that a free electron can again be trapped in the center. Moreover, the rate of filling of states in the conduction band and the rate of filling of traps by conduction electrons are controlled by the system of nonlinear differential equations involved in the kinetics of the charge carriers. A careful analysis of the system of equations is needed to understand the traffic of electrons between the conduction band and the traps (and vice versa), because it is a random process and therefore the electron density in the conduction band fluctuates and the instantaneous value of the lifetime is also altered. This creates a new type of noise, which is discussed below.

5.3.10. Noise Due to the Stability of a System of Differential Equations

Recently, on the basis of the above considerations, an interesting approach was taken by Joshi [57]. Here, the system with one kind of trap and recombination center, which is very common in large-band-gap semiconductors, is discussed. Even though the equations presented here are valid for a system of one type of trap, it can be generalized for other systems by changing or adding a few sets of parameters.

The fundamental mechanisms of the generation and annihilation of the photoexcited charge carriers that produce fluctuations are well understood (they are not a Poisson distribution), and hence a basic theory for a time-dependent process that includes noise is known with reasonable accuracy and some predictions are possible. In the presence of traps and recombination centers the non-steady-state equations governing the photocurrent for electrons and holes are given by [57]

$$\frac{d\underline{n}}{dt} = g_0 - \alpha_1 \underline{n}(N_t - n_c) + \gamma_1 n_c - c_1 \underline{n} \tag{5.45a}$$

$$\frac{dn_c}{dt} = \alpha_1 \underline{n}(N_t - n_c) - \delta_0 n_c \underline{p} - \gamma_1 n_c \tag{5.45b}$$

$$\frac{d\underline{p}}{dt} = g_0 - \delta_0 n_c \underline{p} - c_2 \underline{p} \tag{5.45c}$$

where c_1 and c_2 are the capture cross sections of the recombination centers for electrons and holes, respectively. The other symbols were explained in Chapter 3.* If one assumes that the major contribution in the decay process comes from Eq. (5.45a), and that the effect of the traps is negligible, then the decay curve can be written as

$$\underline{n} = \underline{n}_0 e^{-t/\tau} \tag{5.46}$$

and its behavior near the equilibrium point can be best approximated by Eq. (5.46). However, in the presence of traps, this equation is no longer valid even in an approximate manner near the equilibrium point ($I_{pho} = 0$).

---

*A similar system of equations, with a slight modification, appears in the literature. The conclusions obtained here can be extended to those systems.

The approximate solution obtained in a convenient time domain does not provide useful information, because this is a nonlinear system of equations and the trajectory of the photocurrent at a time near the equilibrium point is determined by the stability of the system. The latter can be examined with the help of phase plane analysis.

The stability of the system can be evaluated by Liapunov's direct method [58,59], according to which the nature of the stability depends on the eigenvalues of the matrix corresponding to the linear part of the present system of differential equations. Eigenvalues for the present system have been obtained,

$$\begin{bmatrix} \underline{n} \\ n_c \\ \underline{p} \end{bmatrix} = \overline{M} \begin{bmatrix} \underline{n} \\ n_c \\ \underline{p} \end{bmatrix} \tag{5.47}$$

where

$$\overline{M} - \begin{bmatrix} -\alpha_1 N_t - c_1 & -\gamma_1 & 0 \\ \alpha_1 N_t & -\gamma_1 & 0 \\ 0 & 0 & 0 \end{bmatrix}$$

Since $|\overline{M} - \lambda\overline{I}| = 0$, eigenvalues of the matrix are [57]

$$\lambda_1 = -c_2$$

$$\lambda_2 = 1/2\{-(N_t\alpha_1 + c_1 + \gamma_1) - [(\alpha_1 N_t + c_1 + \gamma_1)^2 - 4\gamma_1 c_1]^{1/2}$$

$$\lambda_3 = 1/2\{-(\alpha_1 N_t + c_1 + \gamma_1) + [(\alpha_1 N_t + c_1 + \gamma_1)^2 - 4\gamma_1 c_1]^{1/2}$$

All three roots are real and negative, so the present system is asymptotically stable.

When the excess of electrons or holes falls to some critical value near the equilibrium state ($I_{pho} = 0$), the solution goes to the equilibrium point in an undetermined manner. This suggests that the long time limit approach to the equilibrium part of the relaxation curve is not reproducible. Thus, it behaves unpredictably near the equilibrium value. This is a noise in photoconductors caused by the nature of nonlinear differential equations involved in the kinetics of the photoconductivity process.

Physically, it is possible to understand the origin for this type of noise. Electrons from the traps are ejected to the conduction band with thermal energy kT. Similarly, electrons from the conduction band are captured in the traps. The interplay between the systems generates a variable number of free electrons as a function of time; hence, fluctuations in the photocurrent are expected. Details of noise depend on the parameters of the traps and the operating temperature of the system.

It should be possible to obtain information about the fluctuations in the density of electrons in the conduction band and the density of occupied traps and holes in the valence band with the help of three-dimensional phase analysis. However, the mathematical technique for this analysis has not been developed for such a complicated three-dimensional system. Therefore, we consider the fluctuations of electrons in the conduction band and the rate of transfer of electrons in the traps as a function of time. The variation in the number of electrons in the conduction band and the variation in the number of occupied traps can be understood with the help of the following equation [57]:

$$\frac{d\underline{n}}{dn_c} = -\frac{\underline{n}[c_1 + \alpha_1(N_t - n_c)] + \gamma_1 n_c}{\underline{n}[\alpha_1(N_t - n_c)] - \gamma_1 n_c} \tag{5.48}$$

when n_c is small enough,

$$\gamma_1 n_c << \underline{n}[c_1 + \alpha_1(N_t - n_c] \tag{5.49a}$$

and in this region becomes

$$\frac{d\underline{n}}{dn_c} = -\frac{1 + c_3(N_t - n_c)}{c_3(N_t - n_c)} \tag{5.49b}$$

where $c_3 = \alpha_1/c_1$. The solution of this equation is

$$\underline{n} = (N_t - n_c) + \frac{1}{c_3}\ln\,(N_t - n_c) \tag{5.50}$$

The value of the constant of integration is not evaluated as it just shifts the trajectory in the phase diagram and does not modify any argument. Since N_t is a constant and n_c increases as n decreases, this behavior can be understood with the help of Fig. 5.6. This conclusion is against normal expectations, namely, that the numbers of electrons in the traps and in the conduction band should decrease over time. This apparent paradox can be explained easily.

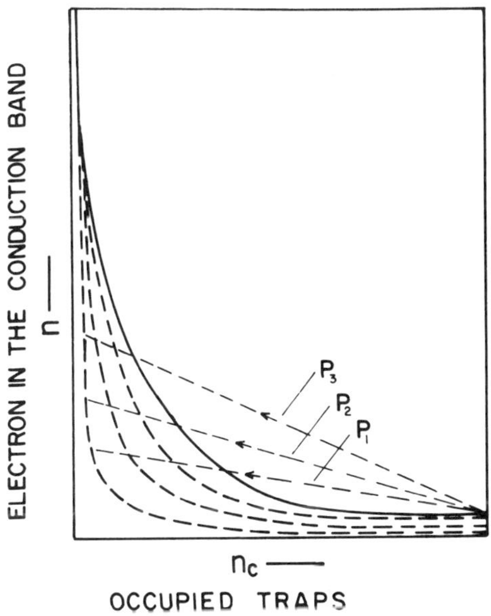

FIGURE 5.6 Phase plane diagram for free electrons and occupied traps [simulated with Eq. (5.55)] for $c_1/\gamma_1 = 1$ (P_1), 0.25 (P_2), 0.5 (P_3). [After Ref. 57.]

As n_c, the number of occupied traps, decreases, the probability of occupation increases noticeably, and hence, again, more traps are occupied. After a certain time, the value of $\underline{n}$ decreases and n_c increases, so that

$$\gamma_1 n_c \gg \alpha_1 \underline{n}(N_t - n_c) \tag{5.51}$$

The equation controlling the trajectory in the phase plane becomes

$$\frac{d\underline{n}}{dn_c} = B\frac{\underline{n}}{n_c} - 1 \tag{5.52}$$

where $B = C_1/\gamma_1$. The solution of this equation is

$$\underline{n} = \frac{n_c^{\ B} + n_c}{B - 1} + K_2 \tag{5.53}$$

where K_2 is a constant of integration and is equal to

$$K_2 = \Delta \underline{n} - \frac{1}{B-1} [n_{co}^B + n_{co}] \quad (5.54)$$

where n_{co} is the number of occupied traps and $\Delta \underline{n}$ is the number of electrons in the conduction band at that time [i.e., when Eq. (5.52) holds true]. Equation (5.54) becomes

$$\underline{n} = \Delta \underline{n} + \frac{1}{B-1} [(n_c^B + n_c) - (n_{co}^B + n_{co}] \quad (5.55)$$

The interplay between electrons in the traps and those in the conduction band is visualized with the help of Fig. 5.6 for various values of B. This creates fluctuations in the charge density in the conduction band, and therefore a new kind of noise is observed [57]. This has often been noticed before photocurrent has reached the steady-state value. The exact magnitude and the range in which such noise will be dominant depend on the parameters of the traps and the difference between the rate at which traps eject electrons to the conduction band and the rate at which they recombine with holes (independent of whether the process is radiative or nonradiative).

A similar approach, considering the noise originated from the nonlinear rate equations, has been adopted by Teitsworth and Westervelt [60] to understand deterministic noise in extrinsic photoconductors. In this case also, the rate equation dealing with charge trapping through acceptor levels is considered along with the variation in the electric field. A similar set of differential equations often forms a two-dimensional system in the form of a damped nonlinear oscillator. Conventional small-signal analysis shows that such a system leads to chaotic oscillations that produce deterministic broad-band noise [61]. Theoretical and experimental studies in this direction have been carried out for far-infrared germanium photoconductors.

The types of noise mentioned above are important and are frequently observed in various types of conventional detectors, but they are not the only types of noise. Recently several new designs and structures have been invented in superlattice or modulation doped semiconductors. It is suspected that miscellaneous types of noise are present in those devices. This aspect will be discussed later.

5.3.11. Importance of Noise in Photodetector Technology

In this chapter, we have discussed some of the commonly observed types of noise and their importance. The relative magnitudes of each depend on several parameters such as the type (n or p) and technology of the semiconductor—metal contact, surface cleanliness, dominant g-r mechanism, lifetime constants, and scattering and recombination processes. Therefore, it is not safe to compare the magnitudes of different types of noise even though they have been compared in the past [45]. As mentioned earlier, each particular device should be studied independently to evaluate noise.

Noise analysis frequently has side advantages. Depending upon its origin, valuable parameters of the system can be extracted. A representative example is the study of deep electron traps. These contribute in low-frequency noise and therefore the study becomes, to a certain extent, a technique complementary to deep-level transient spectroscopy. Using this approach, Loreck et al. [62] estimated deep levels in complicated heterostructures such as the AlGaAs-GaAs system.

Knowledge of the statistics of the excess noise is important, particularly for a digital optical receiver. Recent studies show that in a separate graded band gap and avalanche photodiode, non-Gaussian behavior is observed [60]. No similar study for other types of detectors has yet been reported but considering its importance, it is expected that research work in this direction will be accelerated and such information will be available for other devices.

5.3.12. Preamplifier Considerations

We have seen that the magnitude and spectral information of noise are very useful for detecting low-power radiation. However, they are not sufficient unless low-noise electronics are employed. As mentioned in Chapter 2, noise originating from instruments and matching impedance between the detector circuit and the preamplifier are the crucial factors.

The experimental setup not only creates excess noise but can also alter the noise spectrum of a particular device. In the preamplifier circuit, the detector bias resistance and input resistance of the active element are often shunted by the parallel detector and the input capacitance. An equivalent circuit for Johnson noise is given in Fig. 5.7. It is worth mentioning that the spectral density changes because of the presence of one or more capacitors (a given detector has its capacitance) and is given by the expression [63]

$$\langle V_n^2 \rangle = \frac{4kTR}{1 + (R\omega C_{eff})^2} \qquad (5.56)$$

Such considerations are always necessary.

It makes no sense to develop devices with extremely low noise (of the order of nano- or picoseconds) and not use low-noise elec-

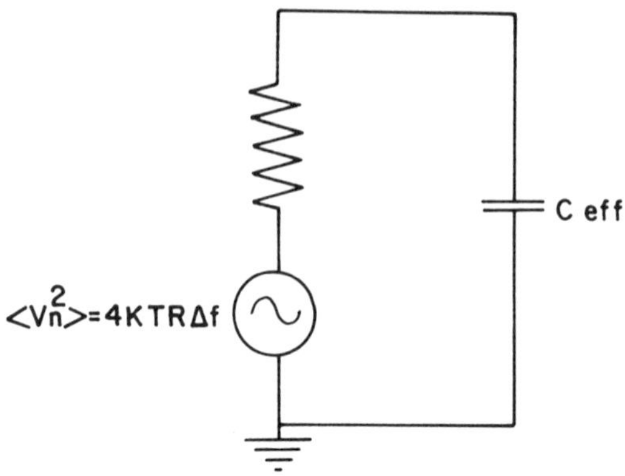

FIGURE 5.7 An equivalent circuit to account for impedance of preamplifier.

tronics. Of course, this is a special topic in electronics, and we will not discuss even the pertinent aspects here. What is important for low-level photodetection is that detector and preamplifier noise should be considered together and due consideration should be given to the experimental setup to minimize it.

REFERENCES

1. A. van der Ziel, Fluctuation Phenomena in Semiconductors. Butterworths, London, 1959.
2. K. M. van Vliet and J. R. Fassett, in Fluctuation Phenomena in Solids. R. E. Burgess, Ed. Academic, New York, 1965.
3. P. Dutta and P. M. Horn, Rev. Mod. Phys. 53,497 (1981).
4. P. W. Kruse, R. D. McGlauchlin, and X. McQuistan, Elements of Infrared Technology. Wiley, New York, 1962.
5. R. Muller, in Noise in Physical Systems. D. Wolf, Ed. Springer-Verlag, New York, 1978, p. 13.
6. R. A. Smith, F. E. Jones, and R. P. Chasmar, The Detection and Measurement of Infrared Radiation. Clarendon Press, Oxford, 1968.
7. R. C. Jones, in Advances in Electronics. L. Marton, Ed. Academic, New York, 1953, p. 2.
8. A. van der Ziel, Noise. Prentice-Hall, Englewood Cliffs, N.J., 1954.
9. K. M. van Vliet, Appl. Opt. 6,1145 (1967).

10. R. H. Kingston, Detection of Optical and Infrared Radiation. Springer-Verlag, Berlin, 1978.
11. W. B. Davenport and W. L. Root, Random Signal and Noise. McGraw-Hill, New York, 1958.
12. J. B. Thomas, An Introduction to Applied Probability and Random Process. Wiley, New York, 1981.
13. P. L. Meyer, Introductory Probability and Statistical Applications. Addison-Wesley, Reading, Mass., 1970.
14. A. Ambroziak, Semiconductor Photoelectric Devices. Iliffe, London, 1968.
15. D. Long and J. Schmit, in Semiconductors and Semimetals, Vol. 5. R. K. Willardson and A. C. Beer, Eds. Academic, New York, 1970, p. 175.
16. P. Kruse, in Semiconductors and Semimetals, Vol. 5, R. K. Willardson and A. C. Beer, Eds. Academic, New York, 1970, p. 15.
17. F. J. Hyde, Semiconductors. Macdonald, London, 1965.
18. S. M. Sze, Physics of Semiconductors Devices. Wiley-Interscience, New York, 1981.
19. J. P. Vilcot, D. Decoster, and L. Raczy, Electron. Lett., 20, 274 (1984).
20. D. L. Smith, J. Appl. Phys., 53,7051 (1982).
21. R. C. Jones, Proc. IRE 47,1481 (1959).
22. R. V. Rollin and I. M. Templeton, Proc. Roy. Phys. Soc. B 66,259 (1953).
23. F. J. Hyde, Proc. Roy. Phys. Soc. B 69,242 (1956).
24. N. V. Joshi, L. Martinez, and R. Echeverria, J. Phys. Chem. Solids 42,281 (1981).
25. A. van der Ziel, Appl. Phys. Lett. 33,883 (1978).
26. A. van der Ziel, Physica 48,242 (1970).
27. A. van der Ziel, Solid State Electron. 25,141 (1982).
28. T. G. M. Kleinpenning, J. Vac. Sci. Technol. A 3,176 (1985).
29. A. van der Ziel, Adv. Electron Phys. 49,225 (1979).
30. T. G. M. Kleinpenning, Physica, 132B:364 (1985).
31. A. van der Ziel, in Noise in Physical Systems. D. Wolf, Ed. Springer-Verlag, New York, 1978, p. 000.
32. S. B. Tobin, S. Iwasa, and T. J. Tredwell, IEEE Trans. Electron. Devices ED-27,43 (1980).
33. H. K. Chung, M. A. Rosenberg, and P. H. Zimmermann, J. Vac. Sci. Technol. A3,189 (1985).
34. J. Bajaj, G. M. Williams, N. H. Sheng, M. Hinnrichs, D. T. Chueung, J. P. Rode, and W. E. Tennant, J. Vac. Sci. Technol. A3,192 (1985).
35. W. A. Redford and C. E. Jones, J. Vac. Sci. Technol. A3,183 (1985).

36. A. H. Verbruggen, R. H. Koch, and C. P. Umbach, Phys. Rev. 35,5864 (1987).
37. J. Kimmerle, W. Kuebart, E. Kuehn, O. Hildebrand, K. Loesch, and G. Seitz in Noise in Physical Systems and 1/f Noise, M. Savelli, G. Lecoy, and J. P. Nougier, Eds. Elsevier, New York, 1983, p. 249.
38. T. G. M. Klienpennig in Noise in Physical Systems and 1/f Noise, M. Savelli, G. Lecoy, and J. P. Nougier, Eds. Elsevier, New York, 1983, p. 285.
39. F. M. Klassen, K. M. vanVliet, and J. Block, Physica, 26:605 (1969).
40. R. J. McIntyre, IEEE Trans. Electron Devices ED-13,164 (1972).
41. K. Kikuchi, T. Okoshi, and A. Hirose, Electron. Lett. 21,801 (1985).
42. A. Antreasyan, C. Y. Chen, and P. A. Garbinski, Electron. Lett. 21,1137 (1985).
43. B. L. Kasper, J. C. Campbell, and A. G. Dental, Electron. Lett. 20,796 (1984).
44. D. D. Coon, R. P. G. Karunasiri, and H. C. Liu, J. Appl. Phys. 60,2636 (1986).
45. A. Rose, Concepts in Photoconductivity, Interscience Tracts on Physics and Astronomy, Vol. 19. Wiley, New York, 1963.
46. G. S. Herzog and A. van der Ziel, Phys. Rev. 84,1249 (1967).
47. J. E. Hill and K. M. van Vliet, J. Appl. Phys. 29,177 (1958).
48. S. Okazaki and M. Hiramatsu, Solid State Electron. 8,401 (1965).
49. P. J. Fellgett, J. Opt. Soc. Am. 39,970 (1949).
50. F. L. Lummis and R. J. Petritz, Phys. Rev. 86,660 (1952).
51. F. M. Klaassen, K. M. van Vliet, and J. Blok, Physica 26,605 (1960).
52. M. A. Carter and J. Woods, J. Phys. D 6,337 (1973).
53. I. E. Türe, M. Claybourn, A. W. Brinkman, and J. Woods, J. Crystal Growth 72,189 (1985).
54. W. Shockley and J. T. Last, Phys. Rev., 107,392 (1957).
55. K. M. van Vliet, Proc. 7th International Congress on Semiconductors, Paris, 1964, p. 831.
56. K. Aoki, K. Miyamae, T. Kobayashi, and K. Yamamoto, J. Phys. Soc. Jpn. 50,357 (1981).
57. N. V. Joshi, Phys. Rev. B 32,1009 (1985).
58. J. P. La Salle, Stability by Liapunov's Direct Method. Academic, New York, 1961.
59. A. L. Rabenstein, Introduction to Ordinary Differential Equations. Academic, New York, 1966.
60. S. M. Teitsworth, and R. M. Westervelt, in Noise in Physical Systems and 1/f Noise, M. Savelli, G. Lecoy, and J. P. Nougier, Eds. Elsevier, New York, 1983, p. 341.

61. P. R. Bratt and R. M. Westervelt, in Semiconductors and Semimetals. R. K. Willardson and A. C. Beer, Eds. Academic, New York, 1977.
62. L. Loreck, H. Dambkes, K. Heime, K. Ploog, and G. Weimann, in Noise in Physical Systems and 1/f Noise, M. Savelli, G. Lecoy, and J. Nougier, Eds., Elsevier, New York, p. 261.
63. R. Benoit, Low noise electronics, in Semiconductor Detectors. G. Bertolini and A. Coche, Eds. American, Elsevier, New York, 1968, p. 201.

6

Modern Photodetectors

6.1. INTRODUCTION

Photodetectors or photosensors have played a crucial role in military science and technology (particularly infrared detectors) for several decades. Their applications are also evident in diverse industries where switching with light intensity is employed. The principle of photoconductivity is used in various devices such as smoke detectors, solar energy cells, sensors, radiation, and electromechanical choppers. There is no limit to the practical applications, and now the importance of photodetectors is increasing because of the proliferation of optical communications. Because of recent technological developments, monolithic integration of ultrahigh-speed photonic devices has become a reality and the design of optical computers is the next logical step.

The variety of advanced applications demands the imposition of stringent requirements regarding performance, ultrahigh speed, detectivity, spectral response, and overall compatibility. It is not easy to meet these requirements, but in the majority of cases modern photodetectors do. Photodetector technology is growing fast to match the needs of the future.

Several comprehensive review papers and books on photodetectors can be found in the literature [1–7]. In this short chapter, no attempt is made to cover all the aspects of this field or all the materials used in photodetectors. Primary emphasis is given to modern photodetectors developed in conjunction with superlattices or quantum well structures. However, in order to achieve a coherent treatment, basic principles and the role of the important parameters involved in the process are also discussed briefly.

6.2. PHYSICAL PRINCIPLES OF n-p JUNCTIONS

The basic theory of joining n and p regions (nondegenerate) in the same semiconductor was first systematically presented in 1949 by Shockley [8]. However, there is evidence that some efforts were made to understand junction properties before this [9,10] and even before the World War II.* The pn junction occurs when a piece of semiconductor has variable concentrations of donors and acceptors in such a way that the transition from donor to acceptor takes place within a reasonably short distance.

Good p-n junctions can be fabricated in several ways. They can be formed in a growth process by adding the proper impurity to the melt. Because of the segregation property, an excess of donors (or acceptors) is observed upon solidification. A p-type impurity can be diffused into an n-type material, leading to a p-n junction. Recent technological advancements in molecular beam epitaxy, metal-organo chemical vapor deposition, liquid-phase epitaxy, and so on makes it possible to introduce an n- or p-type dopant of the desired concentration during the growth of thin films. With the new methods, good control can be maintained, forming reliable, good quality, small physical dimension p-n junctions whose parameters and hence properties can be manipulated to fill a specific need.

There are several interesting books (see, e.g., Refs. 7, 11–13) available on the theory and techniques of p-n junctions. We present here only a brief review of pertinent information on p-n junctions and their application to photodetection.

As mentioned earlier, in an n-p junction the distribution of donors and acceptors varies from one end to other. This means that the distribution of charge carriers also varies. Moreover, the localization of donors and acceptors produces an internal electric field in the semiconductor that favors the separation of photogenerated charge carriers and builds up potential across the junction.

Here we refer to abrupt junctions because not only is the theory much simpler to analyze and the results satisfactory for abrupt junctions but also they illustrate most of the important features of n-p junctions. Moreover they are widely used in device technology. However, it should be kept in mind that a junction obtained by diffusion of donors or acceptors into a semiconductor of the opposite type is generally a graded one. The arguments presented for abrupt junctions are generally applicable to the graded type, although some of the details may vary.

---

*R. S. Ohl at the Bell Telephone Laboratories investigated p-n junctions before World War II.

It was recently found that the devices made from semiconducting alloys with graded concentrations (i.e., graded band gap) show several advantageous features. Throughout our discussion, we will use the term "graded" to refer to the band gap and not to the junction. This will help to avoid confusion between graded junctions and graded band gap material.

It is conventional to refer to an n-p junction when the n-type material has the smaller band gap. Unfortunately, this convention cannot be followed here as we will be using graded band gap and multilayer photodiodes.

A schematic diagram of a p-n junction is shown in Fig. 6.1 along with the concentration of donors and acceptors close to the abrupt junction. When the junction is formed, there is a very high concentration of electrons (holes) on its n (p) side compared to the p (n) side. This creates a concentration gradient, and electrons flow toward the p side. Similarly, holes flow toward the n side. This gives rise to a depletion region (i.e., mobile charges are absent). Positive and negative space charges form the electric dipole layer and build up internal electric fields in such a way as to oppose the flow of electrons from the n region and holes from the p region. In other words, the internal field holds the electron and hole concentrations on the n and p sides, respectively, against the flow tendency caused by the concentration gradient.

The important properties of a p-n junction (not only photoconduction, but many others such as rectification and tunneling) depend on the distribution of the charge density, electric field, and potential near the junction. The form of these quantities can be calculated for an abrupt junction by means of a mathematical model [11,13] that is simple and yet precise enough for modeling sophisticated devices.

6.3. MATHEMATICAL MODEL

We have seen that when the junction is constructed a space charge region is formed in which different charge concentrations are expected as compared to the bulk. We also know that charges are balanced in the bulk, and hence there is not a net positive or negative charge. The charge concentrations in the depletion region can be written as eN_d and eN_a, where N_d and N_a are the donor and acceptor concentrations, respectively. It is a good approximation to assume that charge concentration is a constant on either side of the junction and abruptly becomes zero at the end of the limit of the space charge. To simplify the calculation, it is assumed that charge distribution varies abruptly on both sides of the junction as shown in Fig. 6.2. It is worth mentioning that even

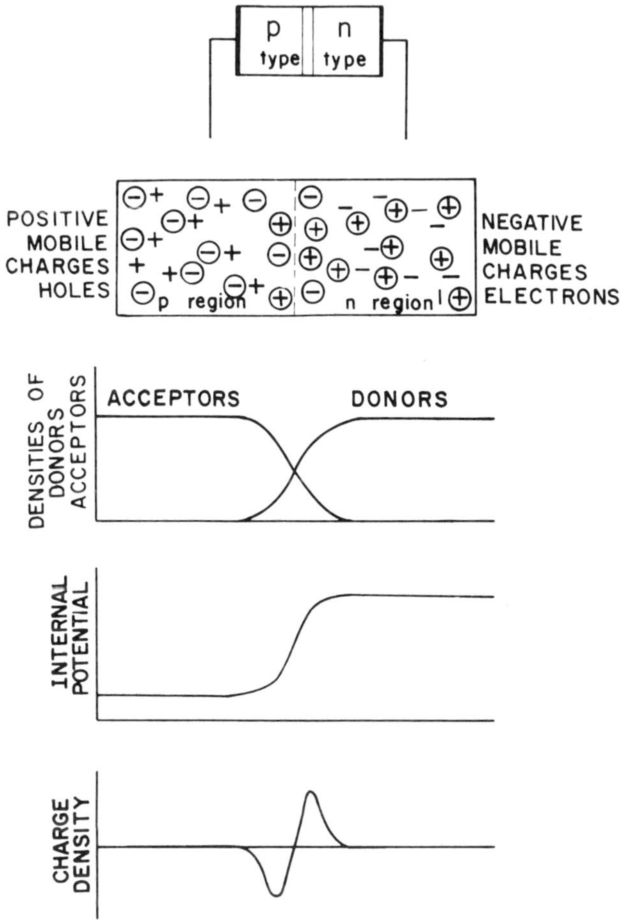

FIGURE 6.1 A p-n junction and formation of a space-charge region. Donor and acceptor concentrations, internal field, and charge density along the p-n junction are also shown.

though the assumption is not completely justified in the true sense, it is acceptable because it explains the observed phenomenon reasonably well.

The variation of charge on either side (p or n) of the junction can be estimated by using Poisson's equation. Electrostatic potential $\underline{\phi}$ as a function of distance x from the junction can be obtained by solving these equations on either side:

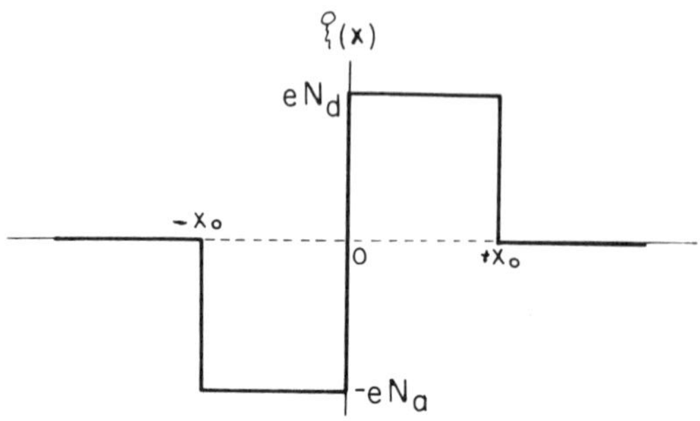

FIGURE 6.2 Charge density for abrupt p-n junction.

$$\frac{d^2 \underline{\phi}}{dx^2} = -\frac{4\pi eN_d}{\varepsilon}, \qquad 0 < x < x_0 \tag{6.1}$$

$$\frac{d^2 \underline{\phi}}{dx^2} = \frac{4\pi eN_a}{\varepsilon}, \qquad -x_0 < x < 0 \tag{6.2}$$

$$\frac{d^2 \underline{\phi}}{dx^2} = 0, \qquad x > x_0, \quad x < -x_0 \tag{6.3}$$

Using appropriate conditions, $\underline{\phi}(x)$ can be written as [13]

$$\underline{\phi}(x) = -\frac{2\pi eN_d}{\varepsilon}(x_0 - x)^2 + C, \qquad 0 < x < x_0 \tag{6.4a}$$

$$= \frac{2\pi eN_a}{\varepsilon}(x + x_0)^2 + D, \qquad -x_0 < x < 0 \tag{6.4b}$$

The only unknown quantity is the width of the depletion region on the n(W_n) and p (W_p) sides of the junction, whose values can easily be estimated if the donor and acceptor concentrations are known, by using the equation expressing the potential difference across the depletion layer, which is given as $N_a W_p = N_d W_p$. The widths W_n and W_p can be written as [13]

$$W_n = \left[\frac{\varepsilon(\underline{\phi}_0 - V)}{2\pi e N_d} \cdot \frac{N_a}{N_d + N_a}\right]^{1/2} \tag{6.5a}$$

$$W_p = \left[\frac{E(\underline{\phi}_0 - V)}{2\pi eN_a} \cdot \frac{N_d}{N_d + N_a}\right]^{1/2} \tag{6.5b}$$

Thus, the total width of the depletion region is given by

$$W = w_n + W_p \tag{6.6a}$$

which becomes

$$W = \left[\frac{\varepsilon(\underline{d} - V)}{2\pi e}\left(\frac{1}{N_a} + \frac{1}{N_d}\right)\right]^{1/2} \tag{6.6b}$$

Equation (6.6b) shows that the depletion width is proportional to the square root of the applied voltage. This will be discussed at a later stage.

It is obvious that nearly all the resistance of a p-n junction comes from the space charge region. Moreover, it behaves as a parallel-plate capacitor whose width increases with the applied reverse bias voltage. Its detailed calculation is given in several publications [e.g., 13].

For designing photodiodes, two important properties of p-n junctions are used. First, electric field is $-d\underline{\phi}/dx$, and one can obtain information on the dependence of E on x. This information is represented in Fig. 6.3. The details of the calculations are available in several textbooks [12,14–17].

The other important property for designing optical detectors is the potential difference between the two sides of the junction. This originates in the process of equalizing the Fermi level on both sides of the junction. Figure 6.4 shows the form of the band gap after it has been formed. It must be realized that the electrons from the p side make use of the advantage of the potential barrier to go down to the n side, and this property is used advantageously

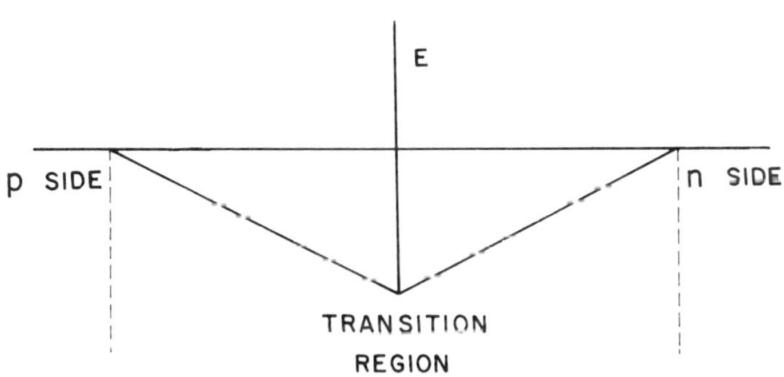

FIGURE 6.3 Electric field obtained by using $E = -\partial\underline{\phi}/\partial x$, where the internal potential ϕ is given in Fig. 6.1.

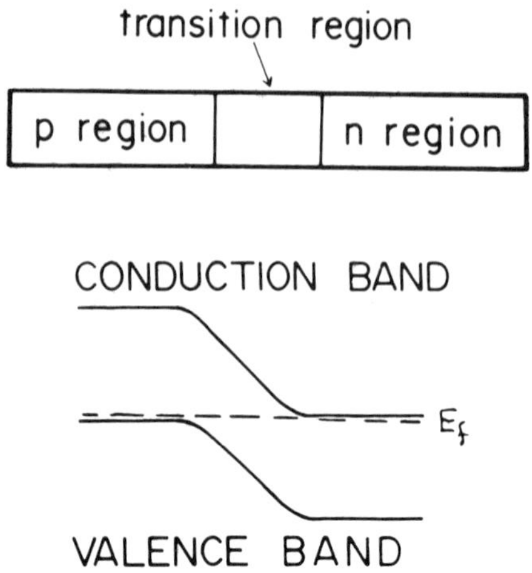

FIGURE 6.4 Potential diagram of p-n junction in equilibrium.

in device technology, particularly in avalanche photodiodes. Figure 6.4 shows the equilibrium situation where no external electric field is applied. The densities of the charge carriers adjust themselves so that there is no net current of electrons or holes in either direction.

Now let us consider the effect of an electric field on a p-n junction.

If a negative potential is applied to a p-type semiconductor, more energy is needed for the electrons to cross the potential barrier, that is, its height is virtually increased. This is called "reverse bias." Even though biasing voltage is a few tenths of a volt, the increase in the barrier height is noticeable (see Fig. 6.5b), and a negligible number of electrons acquire enough thermal energy to surmount the potential barrier. Naturally, electron current from the n region to the p region is reduced dramatically. A similar situation holds for holes. Thus, the flow of the majority carriers almost ceases through the junction. Because the flow of minority carriers is unaffected by the applied field, the current due to them remains unaltered with respect to the voltage, but the total current is substantially reduced.

Now consider the situation when a positive potential is applied to the p side of the semiconductor. Then the barrier height is reduced (Fig. 6.5a), which is called "forward bias." In this case, the potential barriers for electrons is reduced, and a greater number of charge particles can pass it. As a result of this there is an increase in the current. As more voltage is applied, the barrier is further reduced and the flow of charge carriers will continue to

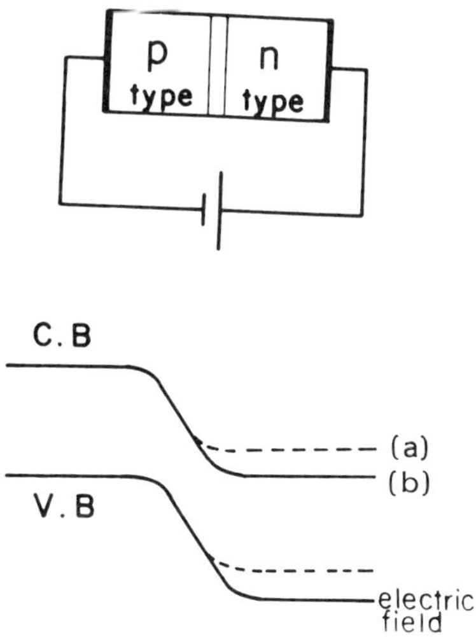

FIGURE 6.5 Potential diagram for p-n junction. (a) Forward bias; (b) reverse bias.

increase. Theoretical calculations of dependence of current (in both the forward and reverse directions) on applied voltage have been carried out and are expressed by [13]

$$I = I_{Rs}[\exp(eV/kT) - 1] \qquad (6.7)$$

where I_{Rs} is the reverse bias saturation current and V is the applied voltage. Typical I—V curves in the forward and backward directions are shown in Fig. 6.6.

6.4. PHOTOVOLTAIC EFFECT

Let us consider that electron—hole pairs are created by the absorption of radiation at or near the p-n junction. The charge carriers formed in the depletion region are immediately separated. The holes are swept toward the p side and electrons toward the n side because of the internal electric field in the depletion region. The charges that are formed within the diffusion length either recombine or diffuse to the boundary of the depletion layer, where they are separated. The minority carriers from both sides are swept across the junction, increasing the positive charge on the p side and the negative charge on the n side. This is equivalent to the forward bias situation. Thus, because of the incident radiation,

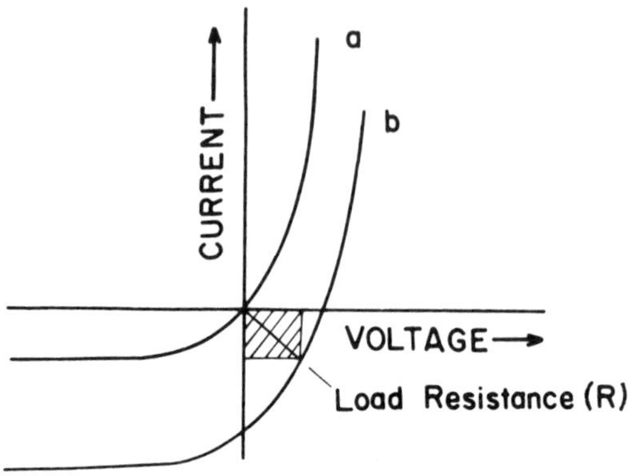

FIGURE 6.6 I—V curves for p-n junction (a) in the dark; (b) under illumination.

bias voltage is built up, which reduces the barrier height and produces forward current (see Fig. 6.7). The magnitude of the forward current is equal to the photocurrent, and the direction is opposite.

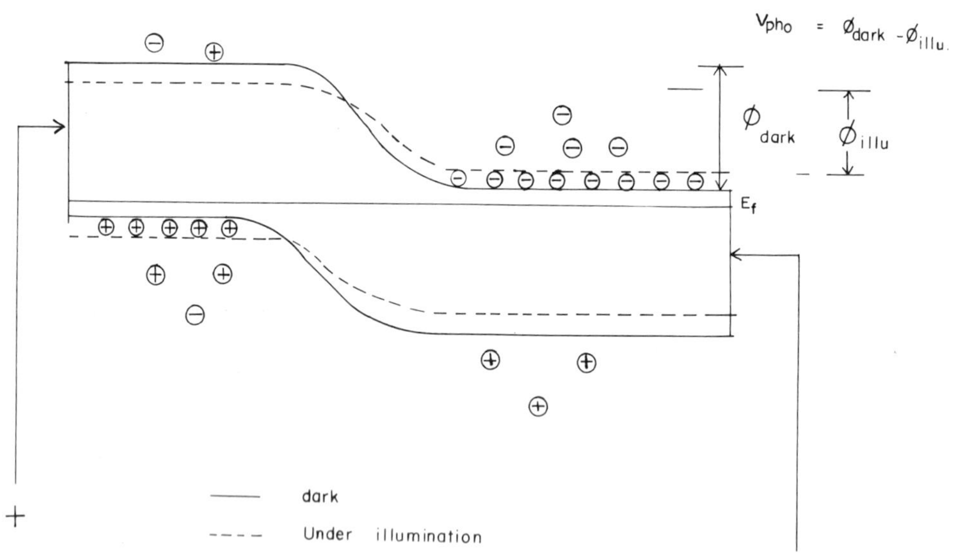

FIGURE 6.7 Effect of the radiation on the barrier height of p-n junction.

The photocurrent can therefore be given as [17,18]

$$-I_{ph} = I_{Rs}[\exp(eV/kT) - 1] \tag{6.8}$$

where I_{Rs} is the reverse saturation current through the junction.

I–V curves of a pn junction under illumination and without illumination are given in Fig. 6.6. Extensive literature is available (see, for example, Refs. 11–18) on this topic, and therefore we skip this discussion here. One important aspect can be visualized—the illumination depresses the I–V characteristic in the fourth quadrant (current negative and voltage positive) and shifts the reverse saturation current toward the more negative side (third quadrant, both current and voltage are negative) rather uniformly; that is, the magnitude of the photocurrent is proportional to the intensity of the radiation. Both of these features have a remarkable importance. The first is useful for the conversion of radiant energy into electric power (solar cells), while the second is used for the detection of radiation by applying negative bias.

As mentioned earlier, diodes can be used as light sensors when they are illuminated by radiation of the proper wavelength and are then called photodiodes.

There are two ways of using a diode as a photosensor. Either the diode is illuminated parallel to the junction plane or it is illuminated perpendicular to the junction. Both configurations are shown in Fig. 6.8. Mathematical analysis [11] shows that for both modes of illumination, the equation of the current–voltage characteristic of the photodiode has the same form [Eq. (6.8)] but the dependence of some of the junction parameters on photocurrent, sensitivity, response time, and so on is entirely different. This aspect deserves due consideration in photodiode design.

The basic difference is not only in the assumptions involved in the two modes but in the fundamental equations that determine the generation of charge carriers and their flow to the electrodes. Assuming that the contribution of the drift current is negligible and that the transition region where the internal electric field is created is also small compared to the p and n regions, the transport of charge carriers is primarily due to the diffusion current (in both modes), and the continuity equations for the minority carriers can be written as [11]

$$\frac{\partial n}{\partial t} = g_n + D_e \frac{\partial^2 n}{\partial x^2} - \frac{n}{\tau_n} \tag{6.9a}$$

In a steady-state condition Eq. (6.9a) becomes

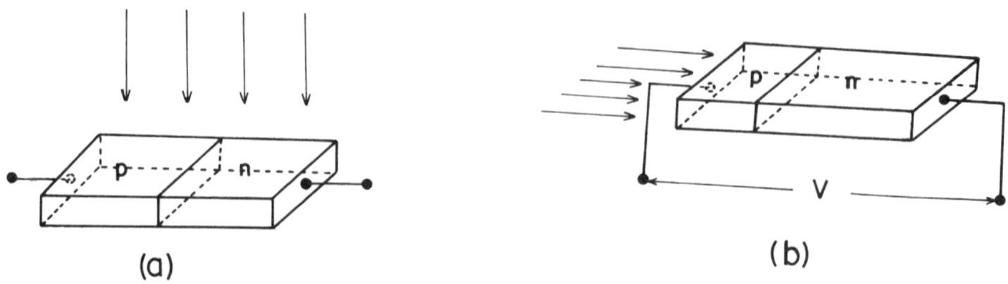

FIGURE 6.8 p-n junction illuminated (a) parallel and (b) perpendicular to the junction plane.

$$D_e \frac{\partial^2 n}{\partial x^2} + g_n - \frac{n}{\tau_n} = 0 \tag{6.9b}$$

The solution of Eq. (6.9b) depends upon the boundary conditions, which are considerably different for the two configurations [11]. This makes a crucial difference in the mathematical expression. Moreover, the fundamental processes such as transport and recombination of charge carriers, which form the structure of the photovoltaic process, are involved in a different manner to build up photovoltage. In addition, there exists an important difference between the two configurations. When the incident radiation is perpendicular to the junction plane, the photogeneration of charge carriers is uniform in the entire p and n regions; meanwhile, in the other case, it is not so. Spatial distribution of photogeneration rate is given by [11]

$$g_n = C_1 e^{-\alpha x} \tag{6.10}$$

Here the constant C_1 includes quantum efficiency and reflectivity.

Obtaining the solution of partial differential equation (6.9b) is tedious and the solution varies with the boundary conditions, which depend on approximations. For certain cases, analytical expressions for sensitivity and other parameters have been worked out [11]. It has been found that in these configurations the I—V curve has the same form but that the magnitude of the saturated photocurrent varies for different configurations.

When the device is illuminated parallel to the junction, the magnitude of the photocurrent is directly proportional to the bulk carrier photogeneration rate and the sum of the diffusion lengths of the carriers on either side of the junction. Photocurrent is also proportional to the illuminated area but not to the thickness as long as the material is thick enough to absorb the radiation. In the case of a device illuminated perpendicular to the junction plane, thickness plays an important role. The front region should be sufficiently thin that photogenerated carriers can reach the potential barrier, while the back region should be thick enough to absorb all the remaining radiation. Charge carriers generated in both regions contribute in the photoconduction process.

Considering the complexity of the solution of Eqs. (6.9), the role of the different parameters involved, and the assumptions and approximations made in both the process and boundary conditions, it is recommended to estimate the properties for a specific device in question rather than to estimate on a general treatment.

A guideline for the calculation of magnitude of photoresponse for various parameters of p-n junction has been provided by Ambroziak [11] for both perpendicular and parallel modes. Photogenerated electrons in the p region are accelerated down the potential hill and reach the n region. Generally, a photodiode is used in the reverse bias mode, and in this case the saturation current arises from the flow of the minority charge carriers and is independent of the applied voltage. If radiation of an appropriate wavelength is incident on the junction, electron—hole pairs are created. Some of these carriers will diffuse through the semiconductor toward the junction. The minority carriers will be swept across the depletion layer by the electric field. The process will increase the saturation current according to the number of photons absorbed. Figure 6.9 shows the effect of radiation on the saturation current in the reverse bias mode. It is worth mentioning that there is usually an excellent linear relationship between photovoltage and intensity of the radiation for five orders of magnitude, and hence photodiodes are suitable for measuring incident power.

Because photogenerated charge carriers build up voltage, the photodiode can be used without bias voltage. This case is called the photovoltaic mode, and the circuit is shown in Fig. 6.10. In this mode, the dark current is minimum, and the response is slow; naturally, this mode is more suitable for low-noise, low-frequency applications. It is very important to remember that shunt resistance and junction capacitance are dependent on the active area. To achieve a higher response, the active area should be increased and the consequences of doing so, such as high circuit resistance and low speed, should be borne in mind.

If the photodiode is to be used as a fast switching device or to detect high-speed light pulses, then it should be operated in the photoconducting mode as shown in Fig. 6.10b. High speed is achieved because the applied field increases the junction field, accelerates electrons/holes to the electrodes, and reduces the transit time. In this mode, the rise time could be as small as picoseconds.

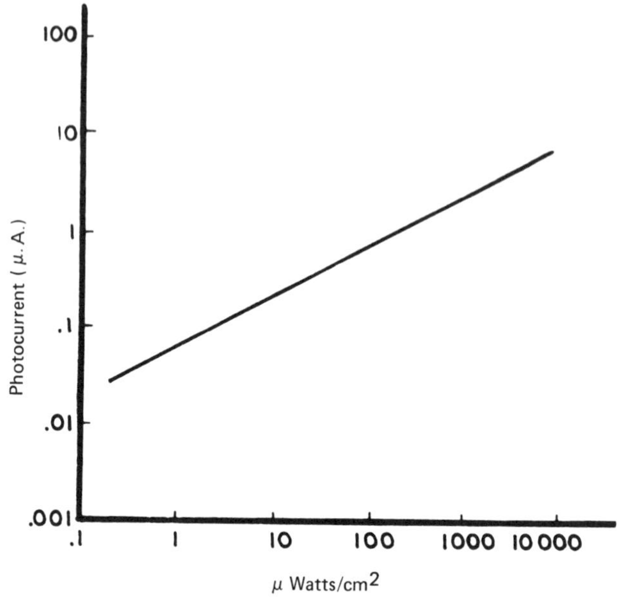

FIGURE 6.9 A typical plot of photocurrent vs. irradiance for a commercial photodiode.

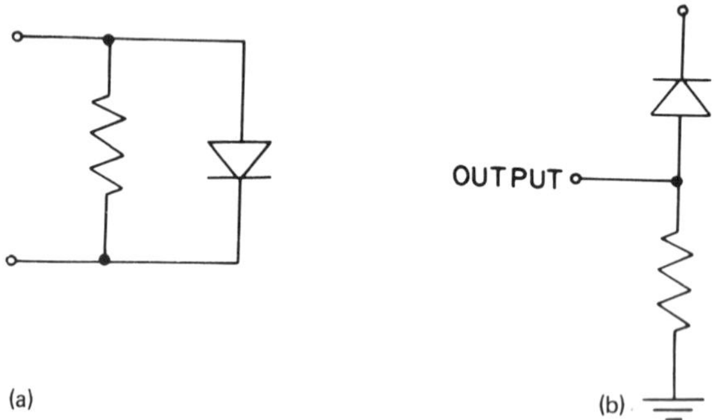

FIGURE 6.10 Photodiode used in (a) the photovoltaic mode and (b) the photoconducting mode.

The s/n ratio is also different in the two modes. In the photovoltaic mode, dark current and noise level are minimum. When operated in the current mode, the magnitude of noise and its spectrum are also different. This means that the detectivity depends upon the operating mode, and both are frequently used according to the specific application.

Spectral response of a photodiode obviously depends on the band gap of the semiconductor, which can be selected by varying the alloy compositions by a suitable method.

Photodiode applications range from elevator sensors to airborne night vision, and their use is being extended dramatically in several fields because of their high sensitivity, speed, small size, low weight, and ease of operation (particularly their low bias voltage compared to the high voltage needed for photomultipliers). Moreover, the spectral response can be tailored to the specific application.

6.5. MODERN PHOTOSENSORS

Modern photodetectors have in their designs very narrow or very thin heterojunctions like in a quantum well or superlattice structure. In detectors with a superlattice structure, the movement of the charge carriers is confined to a direction parallel to the thin layer and electrons behave like a two-dimensional gas. The optical and electrical properties in a two-dimensional system are different from that of three-dimensional solids. The most important property, in this context, is the high mobility of continued charge carriers.

Photosensors consist of photoconducting detectors; several types of photodiodes such as p-i-n, Schottky, Mott, and avalanche photodiodes with various structures (both multilayer and superlattice); and phototransistors. Special structures and geometries are designed to meet diverse needs and requirements. It is not possible to cover all the aspects with their varieties in the brief space allotted here. Instead, we summarize some of the modern types that show superior performance in some respects (speed, sensitivity, low noise, high detectivity, etc.).

6.5.1. Photoconducting Detectors

Photoconducting detectors (PCDs) are generally employed where either speed [20,21] or photosensitivity [22,23] is of fundamental importance. The main advantage of these detectors is the simplicity of the fabrication process (and hence their economy). The present technology permits the use of good quality, materials having high mobility of electrons and holes and therefore the speed of the device is considerably increased. The simplest and yet very effective device was reported by Lawton [24]. With ohmic contacts separated by 0.4 mm on semiinsulating GaAs:Cr, a fast-response (100 ps) PCD was achieved.

Further improvements can be made by minimizing the transit time by reducing the distance between photogenerated charge carriers and the electrodes. Basically, there are two different approaches to making a contact to the device that ensure short transit time; these are shown in Figs. 6.11a and b. In Fig. 6.11a the contacts are planar and the radiation is incident on them, while in Fig. 6.11b the contacts are made on the front and back of the device. When GaAs or its alloys are used, excellent results are reported [21,23]. Using a planar contact structure (see Fig. 6.11a), detectivity as high as 10^{13} cm $Hz^{1/2}/W$ has been reported [23]. This value is comparable to the detectivity obtained from conventional p-i-n and avalanche photodiodes.

6.5.2. Photodiodes

There are several types of photodiodes (p-n, p-i-n, metal—semiconductor, metal-i-n, semiconductor point contact, and others) that have proved to be very suitable for high-speed use. Depending on their design, they may have a response time on the order of a few picoseconds. Commercially available photodiodes are generally p-n,

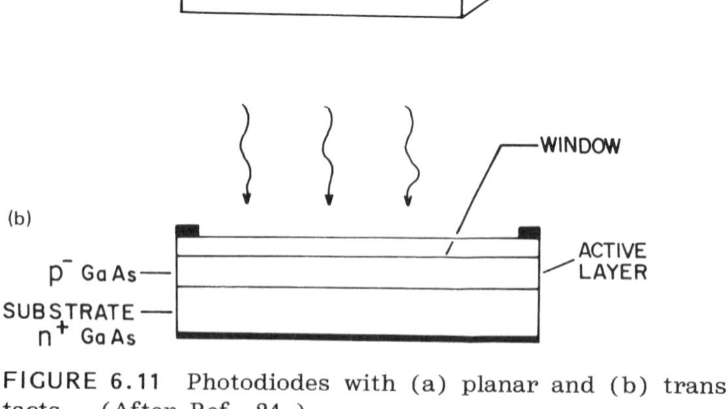

FIGURE 6.11 Photodiodes with (a) planar and (b) transverse contacts. (After Ref. 24.)

p-i-n, or, recently, the avalanche type. Other types such as Schottky and Mott diodes are used for monolithic integration, as their structure is simpler and easy to integrate monolithically on a substrate such as chromium-doped GaAs or iron-doped InP. Avalanche diodes are used for high gain rather than high speed.

a. Ultrahigh-Speed Schottky Photodiode

Schottky-type photodiodes are ideal for high speed as electron–hole pairs are created in the high-field region. This type of diode is becoming very popular because it is easy to fabricate (and hence economical too) and its planar design is suitable for monolithic integration. The best results are demonstrated on high-quality GaAs [25–27] material. A rise time of a few picoseconds is not uncommon for this type of device.

The response speed of the photodiode is controlled by several device parameters but mainly by the transit time between the two electrodes, the time for electron–hole pairs to be swept out from the high-field region. However, other factors such as the RC time constant and parasitic and junction capacitances limit the speed of the device. From Eq. (1.15), it is clear that with a reduction in the value of L, the distance between two electrodes, the transit time is reduced and therefore fast response is manifested in the devices.

In principle, such high-speed photodiodes could be fabricated on any good quality material, but it turns out that almost all the reported data show that GaAs and alloys of GaAs are the best. (In fact, they are the best not only for Schottky-type diodes but for the majority of ultrahigh-speed photonic devices.) This could be attributed to the perfection of GaAs technology, which includes high quality growth, fine control over the diffusion process, sound knowledge of a specific type of contacts, and overall excellent control over processing technology. This is a fine combination of technology and art.

In Schottky-type photodiodes, contacts are not necessarily planar. A typical example of a transverse Schottky photodiode is shown in Fig. 6.12a. It is constructed by growing a thin layer of n^- GaAs on n^+ GaAs by the molecular beam epitaxial method or any other suitable method. A semitransparent platinum film is deposited to make a Schottky barrier contact on n-type GaAs. (There are also alternative methods for making a good and reliable Schottky barrier contact on n-type GaAs.) The photogenerated charges are collected by making ohmic contact on the other side of the substrate.

The working of the photodiode and the reason for its high speed can be understood with the help of the band diagram of the Schottky barriers (see Fig. 6.12b). The radiation of band-gap wavelength is incident on the metal, forming the Schottky barrier. The elec-

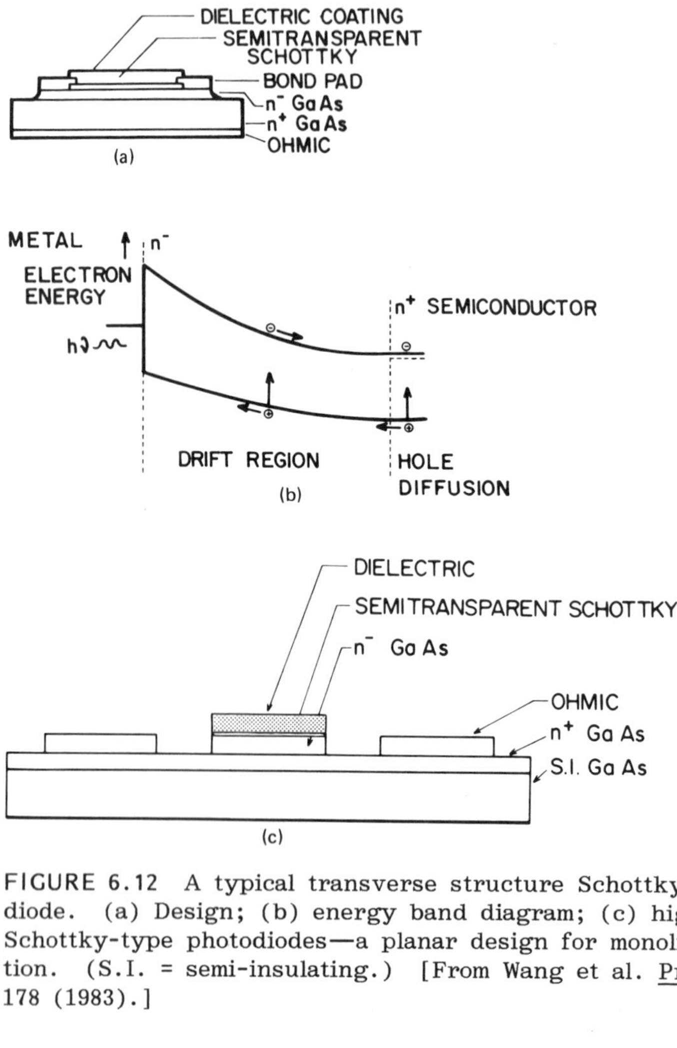

FIGURE 6.12 A typical transverse structure Schottky-type photodiode. (a) Design; (b) energy band diagram; (c) high-speed Schottky-type photodiodes—a planar design for monolithic integration. (S.I. = semi-insulating.) [From Wang et al. Proc. SPIE, 439, 178 (1983).]

tron—hole pairs are created in the depletion layer, which is approximately 1 μm thick, and the charge carriers are immediately separated under reverse bias. For low voltage (about 2—5 V), the junction capacitance could be on the order of 100—150 fF. The structure shown gives very low dark current and also low leakage current. It has been found that such a structure gives up to 20 GHz modulation capability. It was thought that the speed could be further increased by reducing the width of the depletion region and increasing the bias voltage. However, it has been found experimentally that with either method the combination of junction and parasitic capacitance increases, limiting the speed of the detector. This impediment can be overcome by altering the design of the photodiode.

The planar structure shown in Fig. 6.12c is more suitable for higher speed. An n^+-type epitaxial layer is grown on semi-insulating chromium-doped GaAs. On the top of it an n^- layer was grown and semitransparent Schottky contacts were made above it. Ohmic contacts were made on the same side as shown in the figure. As mentioned earlier, GaAs processing technology is well known and with a proper etchant the unwanted part of the layers can be eliminated very easily.

With this and other similar designs, it is possible to reduce the capacitance of the device, and hence higher speed can be achieved. (Recently, successful performance at 100 GHz was reported for this structure.) An additional advantage of a planar structure [26] is that it is more suitable for monolithic integration.

Sometimes the response of a Schottky diode shows a long tail rather than the sharp cutoff that is desired. This tail can be reduced, if necessary, by using a Mott barrier photodiode.

b. *Mott Barrier Photodiode*

The structure of a Mott barrier photodiode is also planar, very simple (both for fabrication and for monolithic integration), and yet effective for achieving a good combination of high-speed performance and moderate noise equivalent power [28].

A thin layer of n-type GaAs is grown on semi-insulating GaAs substrate, and aluminum contacts are made on the same side as shown in Fig. 6.13. The active layer is an n-type epitaxial layer ($N \approx 4 \times 10^{15}$ cm^{-3}) with thickness equal to 0.9 μm. If the thickness of the layer is equal to the Debye length, at higher values of

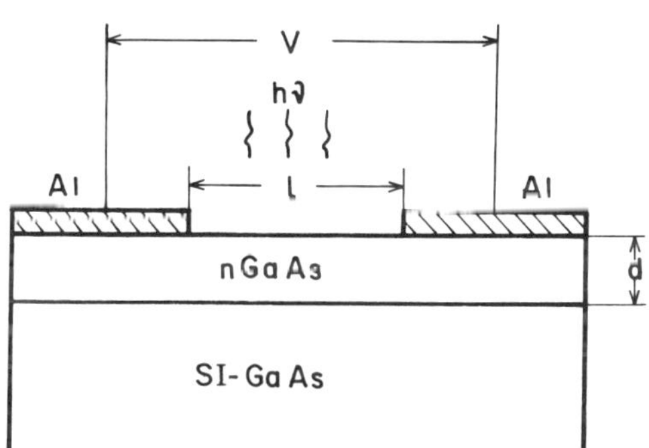

FIGURE 6.13 Mott barrier photodiode. [After Ref. 28.]

applied voltage the device becomes completely depleted. Electron–hole pairs are created in the depleted region, where they are immediately separated.

In such circumstances the device becomes an excellent photodetector. Moreover, the device has an additional advantage, namely low dark current, which is caused by the height of the metal–semiconductor barrier. The important feature of this type of diode is that the capacitance charge time through the external circuit can be reduced considerably.

The reported value of the rise time of such device is 20 ps, the NEP is 4×10^{-11} W, and the gain–bandwidth product is 51 GHz [29]. Thus, the Mott barrier photodiode shows overall an encouraging performance. It is possible to exploit the capabilities of this simple structure for monolithic purposes.

c. *Inverted Heterojunction Photodiode*

In photodetection technology, high speed is not always a fundamental requirement, and a compromise can be made with quantum efficiency, low dark current, or any other device property. A typical example is the inverted heterojunction photodiode, for which quantum efficiency is an important parameter.

The demand for high quantum efficiency comes from defense requirements, particularly for airborne night imaging and for controlling the trajectory of missiles with the help of Nd-YAG infrared lasers (1.06 μm). Recent scientific and technological investigations have proved beyond doubt that heterojunctions of III-V compounds are best suited for devices with high sensitivity, low dark current, low capacitance, and low noise. If the sensitivity is increased by a wise design, then, these attractive features make them unique for the above applications. One of the designs that incorporates these features is the inverted structure.

The design and performance of such a detector (Fig. 6.14) was reported by Eden [29]. Incident radiation penetrates from a high-energy-gap substrate (GaAs is transparent for 1.04 μm) and is absorbed in the depletion region formed between an n^- active layer and a p buffer layer. The band gap of the depletion layer alloy is adjusted by controlling the alloy composition. The absorption coefficient is very high, of the order of 10^4 cm^{-1}. Therefore, almost all the radiation will be absorbed effectively in the depletion region of 1 μm.

If a proper antireflecting coating is used, almost all the incident radiation is absorbed in the depletion region, creating electron–hole pairs and hence, high quantum efficiency ($\simeq$ 100%). Moreover, since the contacts are not at the front, they do not block the incoming radiation.

The reported values of the parameters of the device are excellent; high quantum efficiency, low dark current (3 nA), and low

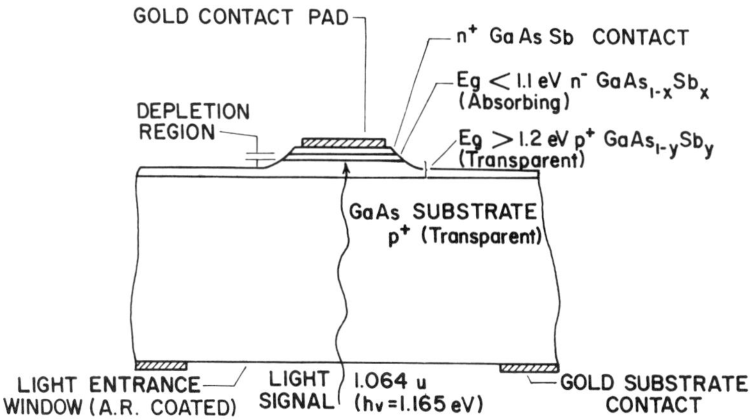

FIGURE 6.14(a) Inverted heterojunction photodiode. [After R. C. Eden. Heterojunction III-V alloy photo-detector for high sensitivity 1.06 μm optical receivers. Proc. IEEE 46, 32, 1975.]

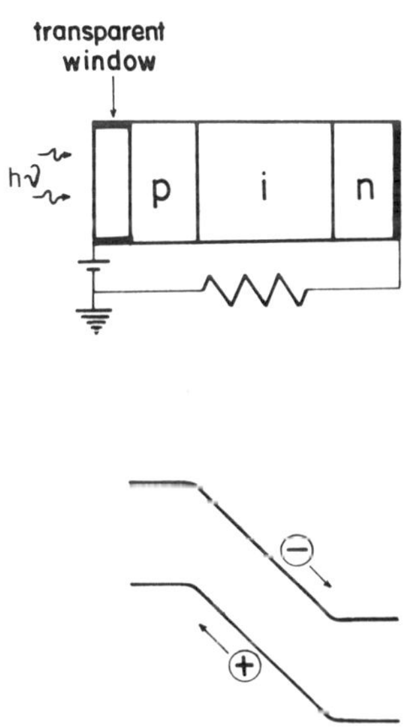

FIGURE 6.14(b) The structure of a conventional p – i – n photodiode and its energy diagram.

capacitance (0.3 pF). The S/N ratio is also better than that of a photomultiplier. Because of its high sensitivity in a very narrow spectral range, this type of photodiode is specifically suitable for laser-illuminated imaging systems.

d. p-i-n Photodiodes

The p-i-n photodiode differs from the conventional p-n diode in that in the former case the p and n regions are separated by a layer of intrinsic semiconductor. In a way, this type of diode consists of two junctions, one n-i and the other p-i, separated by a thin layer. When the diode is operated in the reverse bias mode, the depletion region occupies the i region.

In this design, the width of the depletion region (i region) can be adjusted so that all the incident radiation will be absorbed in this region (width = $1/\alpha$), where the charge carriers are separated effectively. The contribution from electron—hole pairs created outside the depletion region, which generally forms the slow component, is eliminated, and the response in this design is faster. In addition to this, the capacitance of planar p-i-n diodes is lower and helps to increase their modulation capabilities.

There are two ways of using a p-i-n diode [6]. In the first, the front surface is illuminated, while in the second the radiation is incident from the side. The major differences between the two modes are similar to that of p–n photodiodes.

Commercially, p-i-n detectors are available from germanium and silicon, and their rise times are on the order of microseconds. However, in the past few years great progress has been reported [29–33] in the design and construction of these sensors, and considerable improvements have been achieved. The most important is the increase in speed. Planar InGaAs/InP heterostructure p-i-n photodiodes show [30] a response as fast as 40 ps. A similar excellent result has been demonstrated on a hybrid InGaAs p-i-n structure. In this case, a response time on the order of 10 ps was measured [31]. Such high-speed p-i-n diodes are not common, but they are achievable.

The other important property of these photodiodes is their low dark current, which is of the order of a few nanoamperes for commercially available diodes. In this direction also there is remarkable progress. A tendency to exhibit low dark current was shown for InGaAs p-i-n diodes [34,35], and Campbell et al. [32] successfully demonstrated a dark current as low as 50 pA on a planar InGaAs photodiode grown on an n-type InP substrate. This suggests that p-i-n diodes could be proper candidates for fiber communications, where low dark current is one of the fundamental requirements.

The quantum efficiencies of these detectors are reasonably good (about 70%) and depend upon the applied voltage, but an increase in the applied field also increases the dark current, an undesirable property. Fortunately, quantum efficiency and dark current do not both increase in the same way; a typical example of their dependence on the operating voltage is shown in Fig. 6.15b. Quantum efficiency increases rapidly with the field until it attains the saturated value; meanwhile, dark current increases nearly exponentially with the field. For any specific device, the optimum voltage for weak signal detection must be determined experimentally.

6.5.3. Avalanche Photodiode

An analog of a photomultiplier in solid-state photodetectors is the avalanche photodiode, in which the photoexcited carrier produces extra electron–hole pairs as it travels through the space-charge region. These pairs are created by the impact ionization process at about 10^5 V/cm and are characterized by α and β parameters for electrons and holes, respectively. The impact ionization rate is defined as the relative increase in the carrier density per unit length and is expressed as [36]

$$\alpha = \frac{dn}{dx}\frac{1}{n} \tag{6.11}$$

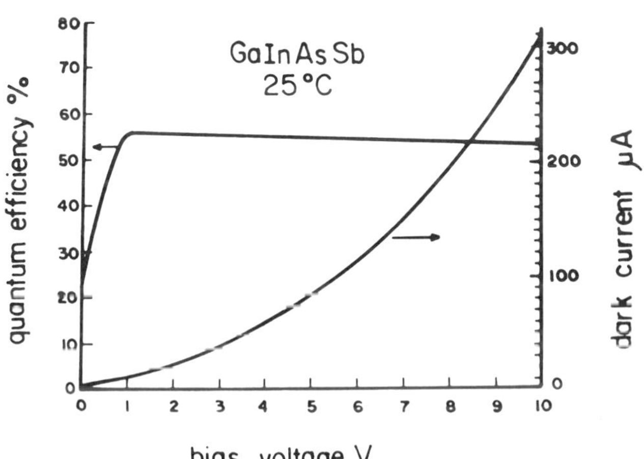

FIGURE 6.15 A plot of quantum efficiency and dark current vs. voltage for a p - i - n photodiode. [After Ref. 33.]

A similar definition holds for holes.

These parameters are strongly electric field dependent and the variation in them is related to the specific material and the range of the electric field. A careful experimental study for a particular device is a fundamental requirement. We will avoid this discussion, and we will limit ourselves here to two simple and practical cases when

$$\beta = 0 \quad \text{and} \quad \alpha = \beta$$

From the definition we can write the equation for the current as a function of the distance in the space-charge region.

$$\frac{dI_n}{dx} = \alpha I_n + \beta I_p \tag{6.12}$$

where x is the distance through which the electron is moved in the depletion region under a high electric field.

First consider the case when $\beta = 0$. Then the current due to the electron is given by

$$I_n(x) = I_0 \exp\left[\int_0^{\alpha} \alpha \, dx\right] \tag{6.13}$$

where $I_n(x)$ is the electron current at x and I_0 is the current at the entrance of the depletion layer. The integral in the exponential cannot be solved analytically because α is not a constant but is a function of the electric field and hence of x. The current due to holes can be obtained from the continuity equation and can be written as

$$I_p(x) = I - I_n(x) \tag{6.14}$$

It can be understood from Eqs. (6.13) and (6.14) that the electron current $I_n(x)$ increases very rapidly as a function of x and in a complicated manner as the exponential of $\int_0^x \alpha \, dx$ increases steeply. Figure 6.16 shows the variation in the electron and hole currents, as a function of distance in the depletion region.

The multiplication factor M is given by the ratio of the current or the charge carrier at the end of the depletion to the corresponding value at the beginning of depletion. When $\alpha = \beta$, this value and the current can be calculated by using Eq. (6.13) and the total current $I = I_n + I_p$.

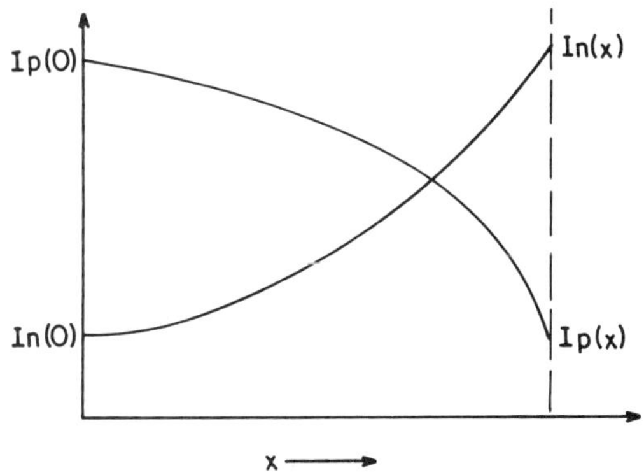

FIGURE 6.16 Electron and hole current originating from the impact-ionization process. [From R. H. Kingston, Detection of Optical and Infrared Radiation, Springer-Verlag, Berlin, 1978.]

$$\frac{dI_n}{dx} = \underline{\alpha} I_n + \underline{\beta}(I - I_n) \tag{6.15}$$

$$= (\underline{\alpha} - \underline{\beta}) I_n + \underline{\beta} I$$

$$= \underline{\alpha} I \tag{6.16}$$

The solution of this equation is

$$I = I_n(W) = \frac{I_{n0}}{1 - \int_0^W \underline{\alpha}\, dx} \tag{6.17}$$

As the width of the depletion region increases, the integral at the denominator, and hence, the current, increases. Of course, it has a limit. When the integral equals 1, breakdown occurs.

The multiplication factor M is a very useful figure that can be obtained with the help of Eq. (6.17) for the special case $\underline{\alpha} = \underline{\beta}$.

$$M \equiv \frac{I_n}{I_{n0}} = \left[1 - \int_0^W \underline{\alpha}\, dx\right]^{-1} \tag{6.18}$$

This shows that for $\int_0^W \underline{\alpha}\, dx = 1$, M is infinite; that is, the avalanche will break down for this voltage, and this should be taken into account for the device applications. Sometimes Eq. (6.18) is used for defining the multiplication factor [37], which cannot be determined without specific knowledge of the variation of the applied field in the depletion region in question and should be obtained experimentally. The other parameter—the thickness of the depletion region—is determined during the fabrication process of the device. Knowledge of both parameters is necessary to design a device with a high multiplication factor.

Some theoretical and experimental works have been reported on the latter aspect and the variation of the ionization coefficients as a function of the electric field for silicon, germanium, and gallium arsenide (both for electrons and for holes); the results are shown in Fig. 6.17 [17]. Theoretical calculations are available due to Baraff [38,39]. It can be seen that the ionization coefficient sharply increases with the applied field, indicating the crucial role of the built-in electric field in the p-n junctions. The relation between the ionization potential and the electric field can be expressed roughly as [37]

$$\underline{\alpha} \propto \exp(-\text{const}/E) \tag{6.19}$$

for germanium and silicon and

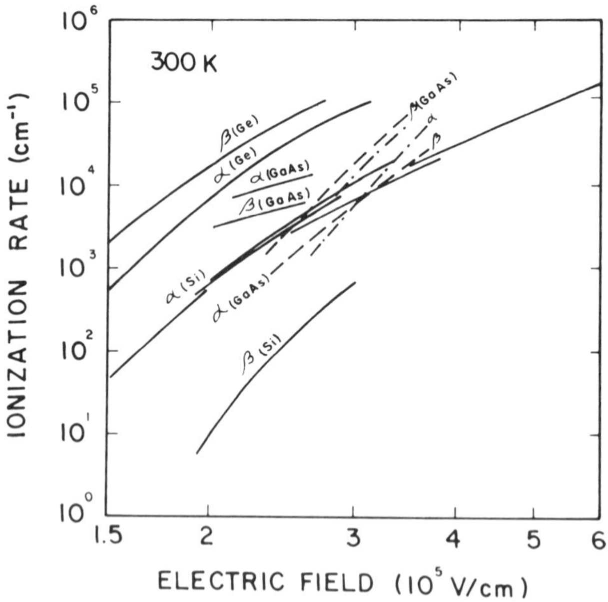

FIGURE 6.17 Impact ionization rate as a function of electric field. [From H. Melchior, Detector and Light Wave Communications.]

$$\alpha \propto \exp(-\text{const}/E^2) \tag{6.20}$$

for gallium arsenide and for gallium phosphate.

The value of the constants in Eqs. (6.19) and (6.20) should be fitted with the experimental data. Needless to say, these constants depend on the temperature and the carrier mean free path, particularly for optical phonon emission. Using a proper voltage and thickness of the depletion layer, a substantial gain in the current has been observed by the impact ionization process.

The multiplication property of the charge carriers in the high-field region has been advantageously used by the device technologist in constructing avalanche photodiodes (APD). The construction of an APD is a little more sophisticated than that of a p-n or p-i-n diode. In addition to the depletion region, where photons are absorbed and primary charge carriers are created, there exists a high-field region where carrier multiplication takes place. This gives internal gain to the device. A good APD should ensure spatial uniformity in the multiplication process by avoiding the region of microplasma, lower voltage breakdown, or excessive leakage.

The presence of the microplasma region and/or a region with a high density of defects, should be avoided by selecting a good quality material and taking care in the processing techniques. In using an APD, either the device is illuminated from the front or it is illuminated from one side. A design and the corresponding electric field variation are shown in Fig. 6.18 to illustrate the region of multiplication. There exist several designs for APDs [40—42] but the principle, namely the multiplication of the charge carriers by impact ionization process, is the same.

Commercially available APDs are fabricated from silicon or germanium and cover the spectral range from the visible to the infrared. Silicon APDs have a cutoff wavelength of 1.1 μm and are not suitable for longer wavelengths for which germanium APDs are used (1.6 μm).

The speed of an APD depends primarily on the transit time of the photogenerated carriers and the capacitance of the device. The transit time depends on the internal field, that is, the doping profile and the multiplication factor M. The transit time $t_{transit}$ is given by [43]

$$t_{transit} = Mt_a + t_d \tag{6.21}$$

where t_a is the transit time of the carriers through the equivalent avalanche region and t_d is the transit time through the equivalent drift region. From Eq. (6.21) it is clear that the price of high gain is an increase in the transit time.

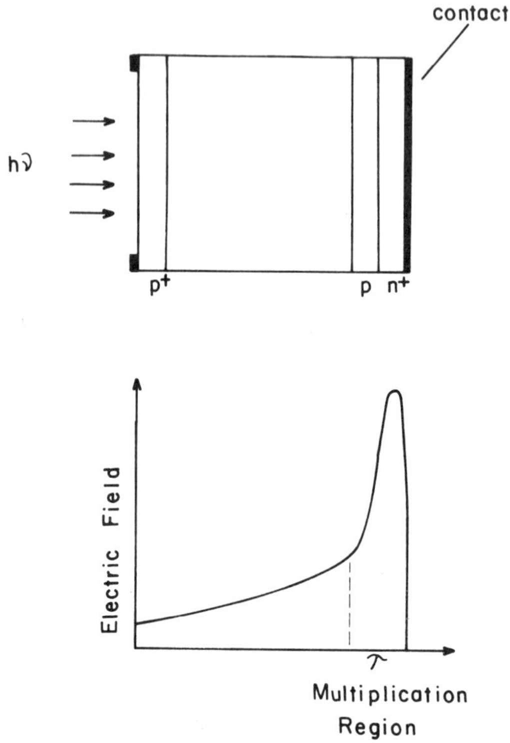

FIGURE 6.18 Top, A typical design of APD. Bottom, Build-up of electric field which causes ionization.

Commercially available APDs generally have a slow response and are not suitable for the measurement of high-speed light pulses. However, recently a few advanced research laboratories have examined III-V alloy heterostructure APDs [43] for high-speed pulse detection, and the results have been encouraging. The rise time of these devices is reported to be less than 35 ps, and the quantum efficiency (taking into account antireflection coating) is about 95%. Good results with other systems such as InGaAsP/InP [44], AlGaAsSb/AlGaSb [45], and AlGaSb/GaSb [46] have also been reported with response times in picoseconds. Today, high-speed APDs are only in the research stage but within a short time they will be available on the market.

The current gain in each device is different and depends on the electric field, which accelerates the electrons, the depletion width, and the geometry of the device. A typical example of the variation in current gain as a function of the reverse bias is shown in Fig. 6.19 [47]. It is clear from this figure that radiation measurement

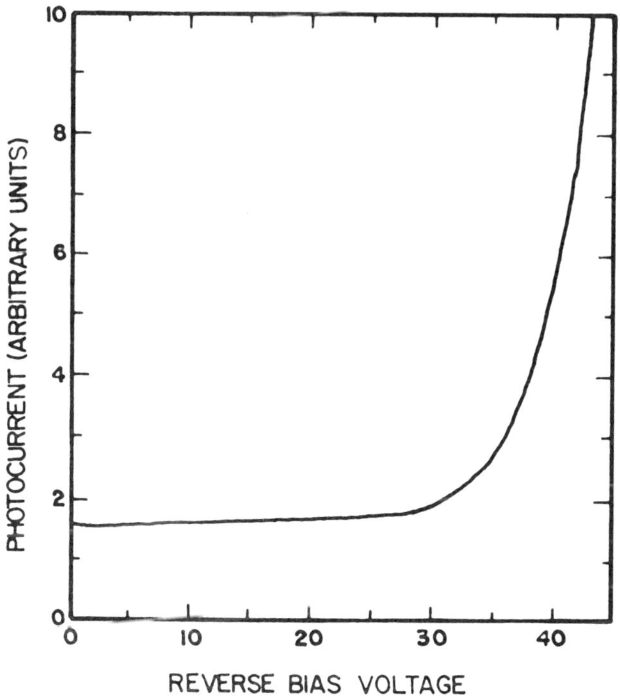

FIGURE 6.19 Photocurrent vs. reverse bias voltage for APD.

with APDs is possible only when the gain is stabilized, and this can be accomplished in several ways, for example, by incorporating an automatic-gain control feedback loop [48].

The noise in APDs has received special attention [6], and it has been found that the ultimate performance depends on the excess noise factor, generally referred to as the F_n factor, which is due to the statistical nature of the avalanche process [49,50]. The F_n factor in turn depends upon the multiplication number M and the ratio of the ionization rate constants α and β. For the electron injection process, F_n is given by [40,51,57]

$$F_n = M_n\left[1 - \left(1 - \frac{\beta}{\alpha}\right)\left(1 - M_n^{-1}\right)^2\right] \tag{6.22}$$

A similar formula holds for the hole injection process. This shows that higher the ratio α/β, the more the F_n factor is reduced. The importance of ratio has been taken into account in the design of APDs.

In avalanche photodiodes, detectivity is generally limited by shot noise, but circuit noise and dark current noise also contribute substantially, and the analysis of noise becomes complicated. Kikuchi et al. [58] carried out a systematic study of shot noise in comparison with circuit and dark current noise on silicon APDs at 77 K. They found that the shot noise can be separated from other types by examining pulse height. Further, it has been shown that by using the proper discrimination circuit, the shot noise can be separated from other types of noise by means of the pulse height analysis technique and applying optimum bias voltage [59]. It is worth mentioning that with this or similar methods, the S/N ratio and detectivity can be improved considerably. Optical power as low as 10^{-17} W has been reported in the literature [58].

Extensive research has been carried out on the AlGaAs/GaAs system, and it has been found that it is an ideal system not only for APDs but also for other optoelectronic devices such as high-speed laser diodes. One of the reasons, among many others is that there is an excellent lattice matching, between alloy AlGaAs and GaAs. This considerably reduces the defects at the interface. Moreover, a high purity material with the desired charge carrier concentration can be grown with relative ease, and the technology for obtaining p-type material by zinc diffusion or another technique is fairly controlled. Above all, the GaAs processing technology is well understood by a few laboratories. This is not to say that other III-V compounds are not used for APDs, but considerable attention has been given to the AlGaAs/GaAs system. For these reasons, for several photonic devices based on superlattice, n-i-p-i and multilayered heterojunction structures the GaAs/$Al_xGa_{1-x}As$ system is preferred.

6.5.4 APD Classifications

It is clear from the above discussion that the impact ionization parameters $\underline{\alpha}$ and $\underline{\beta}$ for electrons and holes play a crucial role in the performance of APDs. According to Eq. (6.22), higher values of the ratio $\underline{\alpha}/\underline{\beta}$ reduce the excess noise and improve the performance of the device. McIntyre [50] has obtained experimental confirmation of this. Several devices based on the increment in the $\underline{\alpha}/\underline{\beta}$ ratio have been reported [51–57,59–65]. They can be classified as follows:

a. Grading the band gap [52,53]
b. Using multilayered heterojunction structure [64]
c. Using both multilayered and graded band gap structures [53]

a. Graded Band Gap APDs

In APDs made from the alloy of GaAs and AlGaAs (Al up to 45%) $\underline{\alpha}$ and $\underline{\beta}$ are nearly equal. One approach to increasing the ratio $\underline{\alpha}/\underline{\beta}$ is to grade the composition of the avalanche region from one end of the depletion layer to the other. This distance is ≈ 0.4 μm [52], but the slight variation from device to device is expected.

As a result of the variation in the composition of the alloy (i.e., a variation in the band gap), a sort of "quasi-electric field" is created that moves both charge carriers toward the lower band-gap region. When a reverse bias is applied to a graded APD, then the electrons are accelerated down the potential hill, that is, they effectively see a higher electric field than the holes, which are moving up the potential hill. The mechanism can be visualized with the help of Fig. 6.20. This helps to increase the ratio $\underline{\alpha}/\underline{\beta}$ significantly. The improved performance of a graded band gap APD is demonstrated by Capasso et al. [52,53].

The impact ionization process in a graded band gap region has been examined by Capasso et al. [53–56]. When a photon is absorbed in this region, an electron–hole pair is created. The electron is accelerated toward the lower band-gap side, and when it acquires sufficient energy an electron–hole pair is created by the impact ionization process. The hole moves to the opposite side (high band-gap side) and also creates an electron–hole pair, but the electrons have a lower ionization energy than the holes because they are accelerated down the potential hill. This increases the $\underline{\alpha}/\underline{\beta}$ ratio, and consequently noise is reduced.

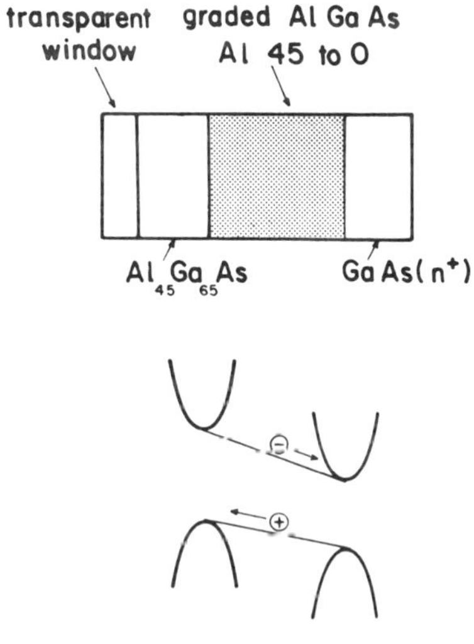

FIGURE 6.20 A conventional APD design and energy band diagram.

Graded APDs have advantages over the nongraded in breakdown behavior also. In the former case, the breakdown of the device is much smoother due to the fact that it starts in the lower band-gap region and then extends to the higher band-gap side, whereas in the nongraded APD the breakdown condition is fulfilled throughout the depletion region. Because of the smoother breakdown, the gain stability of the device is improved.

Instead of making ohmic contacts, Schottky barrier type contacts can also be made. This will improve the collection efficiency, and consequently the responsivity of the device will be further increased. Schottky barrier APDs are not often used but sometimes higher performance has been reported with them. A typical example is the $In_xGa_{1-x}As/GaAs$ system with platinium contacts [60].

The important parameters of graded APDs such as gain, responsivity, response time, multiplication factor, and spectral response depend on the design of the device. These factors can be controlled by the selection of the proper material, composition, and rate of gradation of the constituent elements, thickness of the layers, method of forming p-n junctions, the quality of the compound formed, particularly at the interface (strain and defect levels created), and therefore, the growth technique and, finally, the precision of the technological process. Different laboratories have developed these elements in a variety of ways and so there exist several designs of APDs having more or less similar and excellent performances.

Some compromises are necessary. For example, the multiplication factor can be increased by choosing an adequate gradation, but at the same time the noise will also be increased. Hence, the requirements of the APD must be clear to the designer.

The design of the graded band gap APD has been improved by separating the absorption and multiplication regions [61,62]. A typical example is the InP/InGaAsP/InGaAs structure grown by metalorgano chemical vapor deposition (MOCVD) [55]. In this system, the radiation is absorbed in the alloy $In_xGa_{1-x}As$, whose composition is selected according to the radiation to be detected. InP or its alloy is often used for multiplication purposes. These layers are separated by an intermediate band gap material such as an $In_xGa_{1-x}As_yP_{1-y}$ layer. Such a design shows better performance in terms of dark current and speed. Dark current as low as 10 nanoamperes and speed as high as 100 ps have been demonstrated [61]. A typical structure is shown in Fig. 6.21.

Extensive efforts in technological development of APDs have improved their overall performance, particularly their gain and reliability. Further, when an APD is cooled, the number of thermally generated charge carriers is reduced, and therefore the device can be reverse biased beyond the breakdown voltage corresponding to

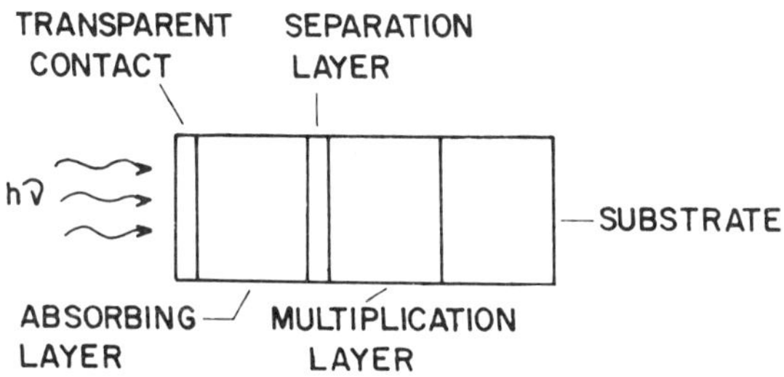

FIGURE 6.21 Avalanche photodiode with separate absorption and multiplication regions.

room temperature. Under these operating conditions, the noise is low and the internal gain is very high for conventional APD, of the order of 10. These are just the right requirements for very low level radiation detection. Further, it has been experimentally confirmed that the detectivity of these devices is superior to that of conventional photomultipliers. Recently Robinson and Metscher [63] detected a single photon with a cooled APD.

b. *Multilayered Heterojunction Structures*

The $\underline{\alpha}$ and $\underline{\beta}$ values in any structure depend exponentially on the impact ionization threshold and on the energy with which charge particles (electrons or holes) are accelerated. If there is a discontinuity in the conduction and valence bands of different magnitude—let us say more in the conduction band than in the valence band—then the ionization rate for electrons will be enhanced and the $\underline{\alpha}/\underline{\beta}$ ratio will be higher [54,64,65].

This was demonstrated by Chin et al. [64] in a multilayer system of $GaAs/Al_xGa_{1-x}As$; the layer thicknesses were adjusted in such a way that at the entrance of every layer of GaAs (low-band-gap material) electrons had sufficient energy to ionize the atom.

Multilayers of a desired composition are generally grown by a suitable technique such as molecular beam epitaxy (MBE) or metal-organo chemical vapor deposition (MOCVD). At the top of the higher-band-gap material, there is an abrupt change to a lower-band-gap alloy. This sudden change in composition creates a step in the band-gap potential, which greatly helps the electrons to acquire enough energy for impact ionization.

c. *Multilayer Graded APD*

The advantages of both the gradation of band gap and the staircase potential obtained by use of a multilayer structure are exploited in this design. The multiplication process can be understood with the help of Fig. 6.22.

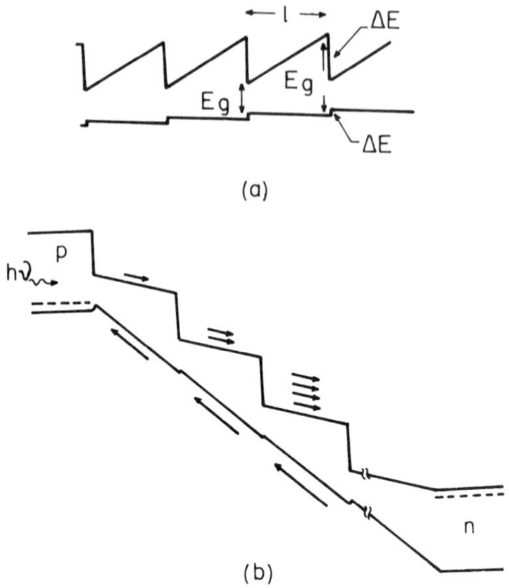

FIGURE 6.22 Multilayer graded band gap APD with characteristic staircase potential design to accelerate impact ionization process. (a) Without electric field; (b) with electric field. [After Ref. 41.]

In such designs, holes cannot have sufficient energy to create electron–hole pairs. The mechanism of charge multiplication can be understood by considering the creation of photoexcited electrons in a p-type material. The excited electrons are accelerated to the n-type material through the graded gap region, in which they do not acquire sufficient energy to create pairs by impact ionization. But at the potential step (entering from the higher-band-gap material to the lower-band-gap region), additional energy is gained, and thus multiplication takes place required for the impact ionization. This happens at successive steps, increasing the number of charge carriers and hence the photoresponse. The process can be compared to that of a vacuum tube, with the potential step corresponding to the dynode.

The main advantage of this design is that a relatively high gain is ensured at low bias voltage. It has been demonstrated that a fabricated multilayered APD of this type can be operated at about 5 V to have an internal gain of 32 [6]. This is indeed a good example of a step APD. In such designs, holes do not get sufficient energy for impact ionization, and the multiplication process is carried out only by electrons and at specific sites rather than in the entire depletion region. For both these reasons, the noise is considerably reduced, and thus the S/N ratio at the output is greatly improved. However, studies are needed to improve the speed of this device and its junction capacitance.

The concepts of multilayered and APD sensors can be extended to superlattice and quantum well structures.

6.5.5. Superlattice and Quantum Well Photosensors

In the last few years substantial improvements in device technology have been obtained through use of superlattice and quantum well structures. As mentioned in Chapter 4, in these structures the motion of electrons is confined to two dimensions and their properties are dramatically altered. In particular, electron/hole mobility is increased remarkably and hence these structures are suitable for high-speed devices such as quantum well sensors and photodiodes (p-n, p-i-n, APD, and multilayer).

a. Superlattice p-i-n Photodiode

In order to achieve low level photodetection, it is customary to use p-i-n photodiode together with field effect transistor preamplifier either monolithically integrated or by hybrid combination. For monolithic integration, a superlattice p-i-n is designed in such a way that the circuit parameters for which better performance (high S/N ratio and low noise level) is observed to match with the impedance of the field effect transistor [66]. The device consists of a GaAs/AlAs p-n junction grown by MBE. The main feature of this diode is its very low dark current ($I_d < 20$ pA) and a response comparable to that of a conventional p-i-n diode; that is, detectivity is increased in this device. The speed (1 GHz) also matches the present performance of devices that are integrated monolithically.

b. Superlattice Avalanche Photodiodes

We have already seen that a key parameter in the design of avalanche photodetectors is the ratio $\underline{\alpha}/\underline{\beta}$, which should be as high as possible. Capasso et al. [68] have demonstrated experimentally that in the superlattice structure of GaAs-AlGaAs this ratio could be as high as 10. This seems contradictory because the impact ionization parameters $\underline{\alpha}$ and $\underline{\beta}$ have the same values in AlGaAs and GaAs bulk materials. The increase in the $\underline{\alpha}/\underline{\beta}$ ratio is attributed to the large difference in the conduction and valence band discontinuities at the $Al_xGa_{1-x}As$/GaAs interface. This is also often the case for other commonly used heterojunctions such as InGaAsP/InP and InAlAs/InGaAs.

In this device a superlattice is grown by alternately depositing layers of AlGaAs (550 Å) and GaAs (450 Å) on n-type GaAs and then depositing p-type GaAs on top of that. A typical device structure and the energy-band diagram corresponding to this structure are shown in Fig. 6.23.

The photoelectron is accelerated in the AlGaAs layer, and when it enters the GaAs layer it acquires additional energy equivalent to the conduction band discontinuity (ΔE_g = 0.48 eV). As α is exponentially proportional to the energy, the effect is remarkable. This argument cannot be extended to the impact parameter β, because the discontinuity in the valence band is only 0.08 eV. Thus, the α/β ratio is increased, giving rise to a high-gain, low-noise device.

Using a similar design, the construction of high-speed superlattice avalanche photodiodes has been reported [69]. The intrinsic response time was as low as 50 ps. A crucial difference in this design was that a slight gradation in the composition was used at the interface, which apparently made the device faster. However, precise information about the transport of the charge carriers in this multiple quantum well (MQW) structure is needed to know what parameters control the speed of these devices and how to make them even faster.

c. Quantum Well Photodetector

This structure is based on a novel concept demonstrated by Coon et al. [70]. It consists of a quantum well formed by an AlGaAs-GaAs-AlGaAs (doped) structure as shown in Fig. 6.24. An incident photon is absorbed by an electron stored in the quantum well, and the radiation is detected through photoemission process. The ener-

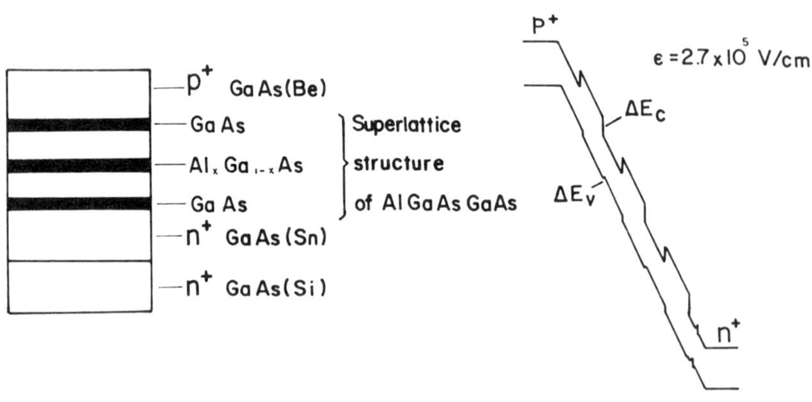

FIGURE 6.23 Structure and energy band diagram of superlattice APD. [After Ref. 41.]

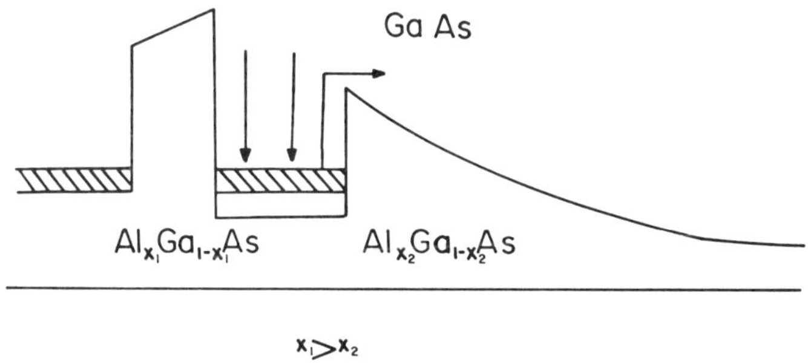

FIGURE 6.24 Quantum well photodetector. [After Ref. 70.]

gy of the photon in this case does not correspond to the band gap of the material but to the energy difference between the occupied state in the quantum well and the top of the barrier. This design eliminates the depletion layer, which is a necessity in conventional photodetectors (p-n, p-i-n, etc.). The schematic band diagram of a biased device is shown in Fig. 6.24. It is important that the Fermi level in the quantum well be equal to the one outside the barrier as shown in this figure.

We know that within the quantum well the energy states of the electron are given by a two-dimensional gas, and the Fermi level can be given by [70]

$$E_{f(2d)} = \left(\frac{\pi\hbar^2 L}{m^*}\right) n \tag{6.23}$$

where L is the length of the well, n is the density of the electrons, and m* is the effective mass of the electron in a two-dimensional potential well. The Fermi level outside the well is given (for three-dimensional solids) by

$$E_{f(3d)} = \left(\frac{\hbar^2}{2m^*}\right) (3\pi^2 n')^{2/3} \tag{6.24}$$

Here, n' is the density of electrons outside the quantum well, which can be adjusted by controlling the impurity concentration so that the Fermi level outside the quantum well matches that inside.

The electrons ejected from the quantum well are replaced by the electrons from the doped region through a resonant tunneling process. The effect of the quantum well parameters on the device performance has been described by Coon et al. [70], and it is found that the potential barrier involved in the photoemission process is the critical parameter for determining the sensitivity of the device.

6.5.6. Doping Heterostructures and Superlattices

Very recently, several designs for high-speed, high-sensitivity photodetectors have been considered. Many of them deal with the heterojunction between a doped high-band-gap semiconductor and low-band-gap material. This is because the modulation doping of semiconductor superlattices and heterojunctions has been shown to alter the transport properties dramatically, particularly at low temperatures [71,72]. A typical example is doped AlGaAs and GaAs structure [73,74]. It is known that modulation doping creates an abrupt discontinuity in the conduction band. Electrons from the donors of wide-band-gap materials (in the present case AlGaAs) transfer across the interface to the lowest conduction band states of lower-band-gap material (GaAs). Thus, the transferred electrons form two-dimensional systems. The electrons are separated from the ionized impurities, and therefore the scattering is reduced, yielding high mobility. It is found that in an ideal heterojunction, mobility varies with $n^{5/2}$, where n is the density of the electrons. Mobility as high as 10^6 $cm^2/V \cdot S$ has been achieved. A dramatic improvement is observed at low temperatures, and hence the performance of the device in which such structure (quantum well, superlattice, and heterostructure) is used, is improved at low temperatures [75]. In addition to this, the detectivity is also improved by increasing the S/N ratio.

A good overview of many aspects of the physics of two-dimensional systems is given in Refs. 76 and 77. The important role of these properties has also been discussed widely. Many of these aspects, therefore, are not covered here. What is important to remember is that devices with such structures show high mobility and hence, fast response.

Making use of these principles, a modulation doped field effect transistor has been used as a radiation detector [78]. A photoconducting detector based on modulation doping has been designed; its performance is excellent, both in speed (15 ps rise time) and in sensitivity (2 A/W) [79].

6.5.7. Photosensor with n-i-p-i Structure

Advantages of two-dimensional properties have been wisely used in designing photosensors with the n-i-p-i structure of PbTe [73,74]. Both the high mobility and the long lifetime of charge carriers, typical characteristics of n-i-p-i structures, help to improve the detectivity of the device. The theoretical limit for the detectivity of photoconductive infrared detectors could apparently be approached. Figure 6.25 shows the reported detectivity and theoretical limit of the photoconductive detector. This is the first experimental data of specific detectivity reported, and slight improvements are expected in the near future.

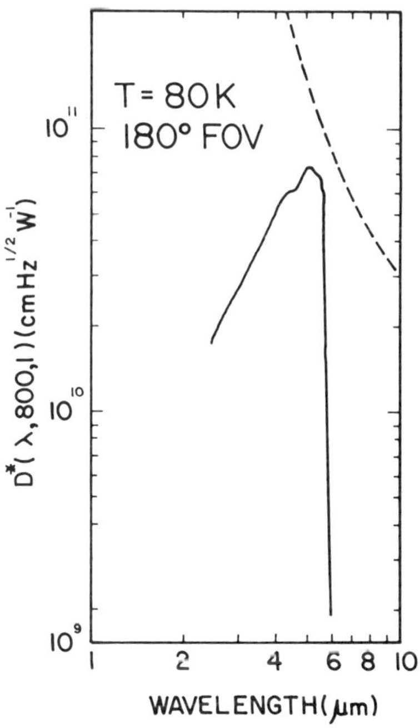

FIGURE 6.25 Experimentally observed detectivity with n-i-p-i structure for PbTe. [After Ref. 73.]

The fundamental concern in designing microstructure photodiodes is the ability to tailor energy band diagrams by using variable-bandgap materials, controlling alloy compositions, selecting proper superlattice structures, and achieving variable doping in them. In the process of modeling band gaps, discontinuities in the valence and conduction bands (not of the same magnitude) are created that accelerate the electrons down the potential hill with the extra energy obtained at the discontinuity. Moreover, the transport and optical properties are dramatically changed in these designs, increasing the speed and sensitivity of the photodiodes. Nature smiles at these devices by providing an excellent lattice match between selected pairs of materials (e.g., GaAs—AlGaAs) and nearly defect-free surfaces. Several aspects of these devices are still being investigated.

6.5.8. Phototransistors

In recent years, no appreciable development, from either the technological or theoretical points of view, has been reported in phototransistors. However, basic information on them is included here for the sake of completeness. We refer only to recently developed designs.

A phototransistor is a light-sensitive active device of moderately high sensitivity and reasonably high speed. Its response, like that of a photodiode, depends upon the wavelength and intensity of the incident radiation, and it behaves like a standard bipolar transistor with an externally controlled collector-base leakage current. Commerically available phototransistors (e.g., Motorola MRD 300 series) cover the complete visible and near-infrared region (450–1005 nm), and are used with a broadband light source such as an incandescent lamp. Their response can be expressed in terms of color temperature. Such phototransistors have several applications in everyday life.

A phototransistor is a device in which the p-n-p (or n-p-n) structure is used and the current increases due to the mobile charges created by the absorption of radiation. A photon flux is allowed to shine on the device in which the electron–hole pairs are generated. A typical p-n-p phototransistor construction is shown in Fig. 6.26 along with the energy band diagram for a biased condition. (In fact, the design is similar to that of conventional transistors except that a transparent window for proper wavelength is provided.) The applied voltage biases the p-n junction in the forward direction so this junction is known as the emitter; the other junction is biased in the reverse direction and is known as the collector.

The way in which this device works is rather simple. The basic principle is essentially the same as that of a p-n junction with some added features. The minority carriers, in the present case holes, diffuse to the collector and contribute to its current. The majority carriers, electrons in the present case, cannot leave the base region and accumulate there only. This increases the space charge at the interface between the base and the emitter, causing a reduction in the potential barrier. Consequently, the flow of holes from emitter to base and further from base to collector also increases. In short, photogenerated charge carriers not only increase the current through the collector junction but also stimulate the flow of charges from the emitter region. The theory for current–voltage characteristics of phototransistors is discussed elsewhere [11,17,23], and simple and rather complete treatments are given by Ambroziak [11] and Dulin [12]. Typical I–V curves between collector and emitter for various intensities of radiation are given in Fig. 6.27.

One of the useful properties of phototransistors is the variation of photocurrent with temperature. It has been found that the photocurrent varies linearly with temperature with a positive slope about 0.7%/°C; because of this, there has recently been a tendency to use this device for monitoring temperature.

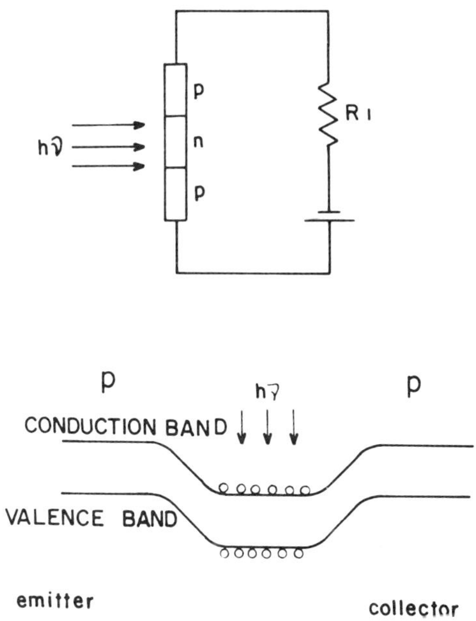

FIGURE 6.26 Top: p-n-p phototransistor. Bottom: its band structure.

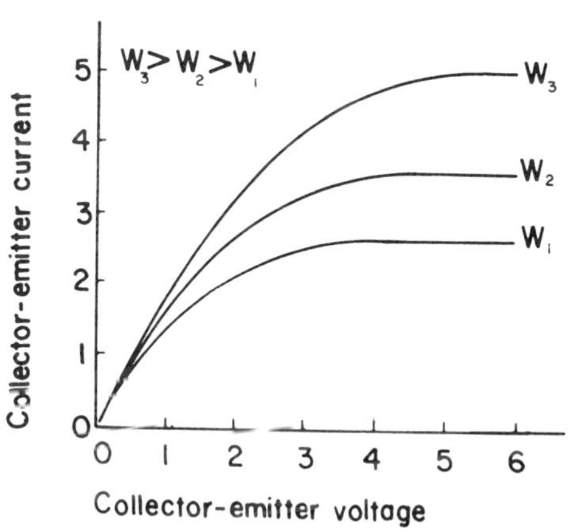

FIGURE 6.27 Typical current–voltage curves for various intensities of incident radiation.

a. Switching Speed

Switching speed is an important parameter of the phototransistor, and its limitations determine the range of applicability. It is known that, unlike photodiodes, the speed of a phototransistor is slow and its rise time is faster than its fall time. The delay in the rise and fall times does not depend upon the quality of the material or the junction formation but can be understood from the equivalent circuit, which is shown in Fig. 6.28 [80], which represents performance reasonably well for a floating base phototransistor. Here G_0 is the forward transconductance, I_c is the collector current, R_{be} is the effective base emitter, resistance, I_{pho} is the photocurrent generated directly by absorption of photons, and C_c, C_e are the capacitance corresponding to the collector and emitter regions, respectively. Circuit analysis shows that a fast rise in the photocurrent I_{pho} will not produce an instantaneous increase in the collector—emitter current. The initial flow of current will be used to charge the capacitor between collector and base and between base and emitter. After charging these capacitors the current will flow through resistance R_{be}. At the end, the generator $g_m V_{be}$ will start operating. A similar situation holds for the fall of photocurrent. Even though photocurrent goes to zero, the discharge of the capacitors through R_{be} will maintain a current flow through the circuit for a considerable amount of time.

It is clear from this discussion that the crucial role in the delay process is played by the capacitance, whose value varies with the applied voltage and is generally within the range of 2—10 pF. A

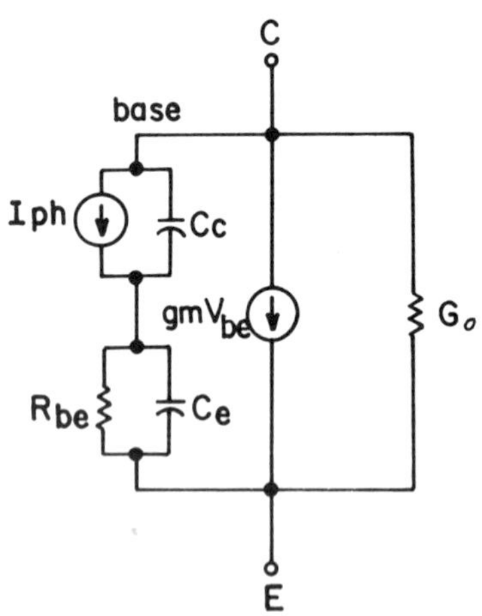

FIGURE 6.28 Equivalent circuit for phototransistor.

knowledge of the dynamic characteristics of phototransistors is necessary when switching speed is of fundamental importance. In short, turn-on and turn-off delays are limited by the capacitance in an equivalent circuit.

There has been an increasing effort to obtain high-gain, high-speed phototransistors. Chand et al. [81] reported a novel phototransistor design to achieve a high-speed (100 ps rise time) that is not possible with traditional methods.

The major impediment to obtaining high speed is the capacitance, whose value can be reduced significantly by reducing the physical dimensions of the device (see Fig. 6.29). Chand et al. [81] successfully reduced the dimensions down to the order of 50 μm. Moreover, in their design the radiation is absorbed completely from the back of the transistor, avoiding the compromise between a narrow base for high gain and a thicker base for photogenerated charge carriers. In short, the present ingenious design increases the speed through small dimensions and at the same time increases gain through efficient absorption of photons.

The design and the processing details have been reported elsewhere [81]. A similar design with slight modifications has been developed on semi-insulating InP [82] that is adequate for monolithic integration. It is expected that such transistors will soon be commercially available.

Up to now, we have discussed several types of modern photodetectors whose performance (high gain, high speed, wider photoresponse, low noise, etc.) is confirmed in the laboratory. Many of these have already been used in industry for research and develop-

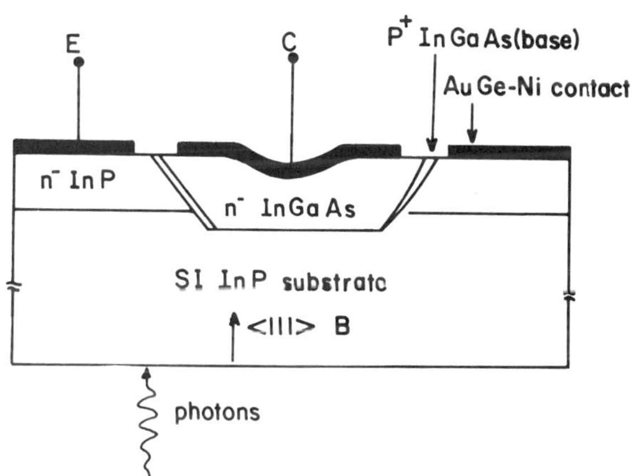

FIGURE 6.29 Design for high-speed phototransistor. [After Ref. 81.]

ment and fiber-optical communication. To complete this topic, we will mention a design for a speculative detector. Even though the performance of such a detector has never been reported, it has some inherent potential advantages.

6.6. A DESIGN FOR A SPECULATIVE DETECTOR

In this section we discuss a novel type of geometry for infrared photodetectors. This structure was proposed by Szmulowicz and Chandler [83] in 1981, but experimental confirmation has not yet been obtained.

According to these authors it is possible to improve overall the performance parameters of the detector such as detectivity, speed, gain, and uniformity by changing its geometry. Instead of the conventional rectangular geometry, they recommend a cylindrical geometry. The electric field then varies with radial distance and hence, the responsivity changes. As we know, responsivity is a key parameter, and an improvement in it directly improves the detectivity and other parameters of the device. We will therefore go quickly through a few mathematical details to get a rough idea of the magnitudes of the proposed improvements and the role of the different parameters in the new design.

Figure 6.30 shows a cross section of the proposed cylindrical detector. As shown in this figure, the radiation is incident perpendicular to the surface. The responsivity of the detector is given by

$$R = I_{pho}/P \tag{6.25}$$

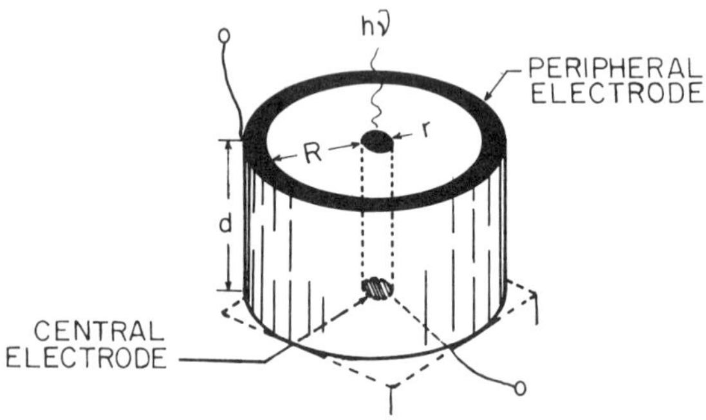

FIGURE 6.30 Cylindrical geometry for photodetectors. [From Ref. 33, "A new cylindrical geometry for transverse infrared photodetectors" IEEE trans. on electron devices. ED-28, 772, 1981. Copyright © 1981 IEEE. Used by permission.]

where I_{pho} is the photocurrent and P is the power of the incident radiation. In this case the total current is

$$I_{tot} = IA \tag{6.26}$$

Here I is the current density and A is the area through which the current passes and is equal to $2\pi rd$. R' is the outer radius.

$$I(r) = \sigma E(r) \tag{6.27a}$$

where

$$E(r) = \frac{1}{r}\left(\frac{V}{\ln(R'/r_0)}\right) \tag{6.27b}$$

The photoresponse (in amperes) is given by

$$I(s) = \frac{2\pi\sigma d}{\ln(R'/r_0)} V \tag{6.28}$$

If ϕ_s is the flux per unit area and d is the thickness of the area in which the radiation is absorbed, then the number of electrons created by the absorption of the radiation is

$$P = \pi R'^2 \phi_s h\nu$$

and

$$\sigma = e\tau_n \mu\eta' \phi_s/d$$

The responsivity, therefore, can be written as

$$R = \frac{e\lambda}{hc}\eta\left[\frac{2\mu V\tau_n}{R'^2 \ln(R'/r_0)}\right] \tag{6.29a}$$

For the rectangular slab, the responsivity is given by

$$R = \frac{e\lambda}{hc}\eta\left[\frac{\mu V\tau_n}{\ell^2}\right] \tag{6.29b}$$

where ℓ is the distance between two electrodes. Comparing Eqs. (6.29a) and (6.29b), it can be seen that the response is improved with cylindrical design. Other performance parameters such as gain, transit time, and detectivity are related to the responsivity and can be evaluated. Detailed calculations of these parameters are given by Szmulowicz and Chandler [83]. What is not evaluated is the thermal radial gradient and its effects on the transport properties. This consideration might alter the predicted values of the performance parameters.

A cylindrical design for photodetectors seems to be attractive and warrants experimental investigation.

6.7. POSITION-SENSITIVE DETECTOR

We have discussed several designs that are used to detect the magnitude of incident radiation. Recently, by using an ingenious method and with a slight modification in the design photoconductive detector, the site of the spot of incident radiation has been located. Such detectors are known as position-sensitive detectors (PSDs).

The design is similar to the previously mentioned detectors. It consists of three layers as shown in Fig. 6.31. The p-type layer is photosensitive and is extremely homogeneous. At both ends of the p layer there are contacts to record output. The back of this layer is coated with n-type material, and at the center an ohmic contact is made to bias the device.

The working principle is very simple. When a light spot falls on a PSD surface, an electric signal proportional to the light intensity is generated and will propagate toward both electrodes. As the resistivity of the p layer is uniform, the magnitude of the photocurrent collected by an electrode is inversely proportional to the dis-

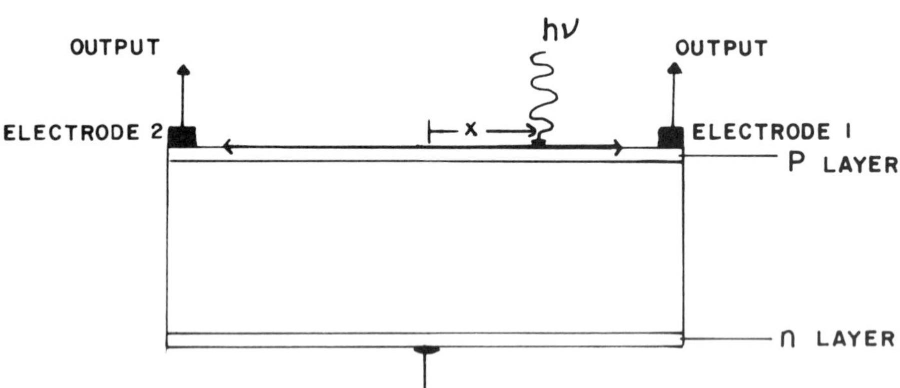

FIGURE 6.31 A design for a position-sensitive detector.

tance between the electrode and the position of the spot. This property has been used to determine the location of the spot as follows.

Assuming that radiation is shining closer to electrode 1 than to electrode 2, the magnitudes of the photocurrent at electrodes 1 and 2 are given by

$$I_1 = I_{tot}C_1/(L - x) \tag{6.30}$$

and

$$I_2 = I_{tot}C_1/(L + x) \tag{6.31}$$

where x is the distance measured from the center, 2L is the total length of the detector, and C_1 is the proportionality constant. (See Fig. 6.31.)

Dividing (6.30) by (6.31) we get

$$\frac{I_1}{I_2} = \frac{L + x}{L - x} \tag{6.32}$$

Remembering that $I_1 + I_2 = I_{tot}$, the output at electrodes 1 and 2 is given by

$$I_1 = I_{tot} \frac{L + x}{2L} \tag{6.33a}$$

and

$$I_2 = I_{tot} \frac{L - x}{2L} \tag{6.33b}$$

Since I_1, I_2, I_{tot}, and L are known, the position of the spot can be estimated with precision.

Position detectors have several applications. They are generally used for autofocusing and laser beam alignment. The possibility of their application in robot vision has widened the perspective of PSDs.

6.8. NOISE IN DETECTORS AND DETECTIVITY

A properly constructed device has very low noise. However, it is necessary to know the magnitude, spectral features, and sources that generate the noise. This information not only helps to reduce the noise but also provides information for improving the device structure and developing optimum working parameters (e.g., voltage, modulating frequency).

The magnitude of contact noise can be reduced considerably by selecting the proper material and methods of making an adequate contact. One of the major sources of noise, particularly in infrared detectors, is 1/f noise. Naturally, for low-signal detection, a higher modulation frequency (10^2 to 10^4 Hz) is recommended.

For certain detectors, the S/N ratio also improves as applied voltage increases, as signal and noise do not increase with voltage in the same proportion. Signal increases at a slightly faster rate than noise; this is illustrated by Fig. 6.32.

Contact and 1/f noise are not the only limiting types of noise observed in detectors. For example, in p-n junction devices, the detection is limited by shot noise [3,17], which is a function of net junction current, and hence this value should be used for evaluating NEP and specific detectivity. Calculated data are available for several device structures [84].

In some types of detectors, it is not always possible to know the limiting type of noise because it depends on which type of recombination process is dominant. Moreover, in several cases there are two or more types of noise that determine detectivity. For example, in

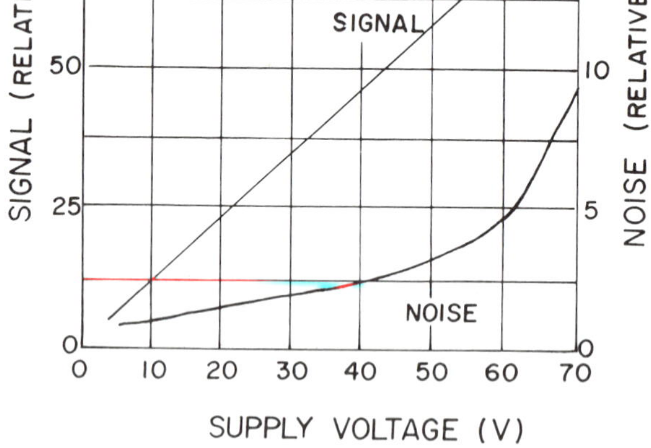

FIGURE 6.32 Dependence of signal and noise on applied voltage.

general, Johnson noise and generation-recombination noise limit the measurement capability of photoconductive detectors. Because these add in quadrature, the total noise to be considered [85,86] is given by

$$V_{tot}^2 = V_{Johnson}^2 + V_{g-r}^2 \tag{6.34}$$

where $V_{Johnson}$ and V_{g-r} are Johnson and generation-recombination noise voltages. As mentioned in Chapter 5, a separate study of g-r noise for a particular device is needed to know which type of recombination process (through traps or band-to-band, etc.) is dominant. In this case only, Eq. (6.34) is used to estimate the limiting noise.

The above discussion creates the impression that for a certain type of device a specific type of noise is dominant, but this is not so. The noise depends on the device design and on the details of processing used in its fabrication. We cite a few examples to illustrate this point.

1. In a detector made through an ion-implantation process or in which a double-layer epitaxial structure is used, there are a considerable number of surface defects, which favor the recombination process. Depending upon the origin, distribution of surface states, p-n junction area, applied voltage and so on, the recombination mechanism and hence the magnitude and noise spectrum vary.
2. In a modulation-doped AlGaAs-GaAs photoconductive detector, noise due to electron temperature has been reported. A similar noise pattern due to hot electrons has been mentioned for photoconducting materials such as silicon [87] and InP [88]. The magnitude of this type of noise increases with applied voltage and finally saturates.
3. Leakage noise is predominant in some photodetectors where proper surface treatment is not possible because of the design or where a guard ring structure is inconvenient. This is typical for monolithically integrated photodetectors, and in the majority of cases the leakage noise can be reduced considerably by appropriate design.

A careful noise analysis using proper statistical methods is certainly a rewarding task not only from the academic point of view but also from the technological point of view as it provides information directly useful for improving the S/N ratio. A typical example is the analysis of dark current that has been carried out for modern photodetectors.

Antreasyan et al. [89] have examined dark current for a planar, interdigitated $Ga_xIn_{1-x}As$ photoconductive device. The structure of the device used in this investigation was very simple and is a widely used design. Therefore, the results could be extended to other similar detectors. In addition to this, similar designs are generally employed in optical signal analysis where incident power is very weak. In such circumstances, dark current analysis could inform the precise role of some parameters in the noise process. Experimental investigation consists in the analysis of the statistical distribution of dark current noise for various bias voltages [90]. Results show that the probability of distribution of dark current about the dc level is Gaussian. Further study shows that the standard deviation of the probability distribution increases with the applied electric field. This means that at low bias voltage both noise and photoresponse are low. But as the voltage increases, both the photoresponse and noise increase at different rates, so there is an optimum bias voltage, which can be estimated. Thus, noise analysis helps to determine an important parameter for low-power radiation measurement.

Dark current analysis for recently developed quantum well photodetectors has also been carried out [91] and it has been concluded that for such types of devices, as expected, the contribution from shot noise is more dominant than that of noise from the thermionic emission process. Extensive investigation in this direction is needed, as modern photodetectors are likely to play a decisive role in long-distance, repeaterless fiber guide communication for transatlantic systems.

For designing and selecting an adequate photosensor and using it for low-level radiation detection, a mere concept of these parameters is not enough. Greater theoretical understanding is necessary.

Here we have given an overview of the modern photodetectors that have been developed within only the past decade. It is clear that technological breakthroughs have permitted considerable increases in responsivity and speed. Some of the devices are available commercially, and others will soon be on the market. A systematic study of the noise and transport properties of two-dimensional devices is needed to improve the overall performance of these devices. Definitely, in coming years we will witness a great development in these aspects.

REFERENCES

1. L. K. Anderson and B. J. McMurtry, Proc. IEEE 54,1335 (1966).
2. L. K. Anderson, J. DiDomenico, Jr., and M. B. Fisher, Adv. Microwaves 5,1 (1970).

3. G. E. Stillman and C. M. Wolfe, Avalanche photodiode, in Semiconductors and Semimetals. R. K. Willardson and A. C. Beer, Eds. Vol. 12, Academic, New York, 1977, p. 291.
4. H. Melchior, J. Lumin. 7,390 (1973).
5. H. Melchior, Phys. Today 30,32 (1977).
6. T. P. Lee and T. Li, Photodetectors, in Optical Fiber Communication. S. E. Miller and Alan G. Chynoweth, Eds. Academic, New York, 1979, p. 593.
7. B. L. Sharma and R. K. Purohit, Semiconductor Heterojunctions. Pergamon, Oxford, 1974.
8. W. Shockley, Bell Syst. Tech. J. 28,435 (1949).
9. L. Sosnowski, J. Starkieichz, and O. Simpson, Nature 159, 818 (1947).
10. S. Benzer, Phys. Rev. 72, 1267 (1947).
11. A. Ambroziak, Semiconductor Photoelectric Devices. Iliffe, London, 1968.
12. V. Dulin, Electron Devices, MIR, Moscow, 1980.
13. J. P. McKelvey, Solid State and Semiconductor Physics. Harper & Row, New York, 1966.
14. J. N. Shive, The Properties, Physics and Design of Semiconductor Devices. Princeton Univ. Press, Princeton, N.J., 1959.
15. W. Shockley, Electrons and Holes in Semiconductors. Van Nostrand, New York, 1950.
16. J. Mulvey, Semiconductor Devices Measurement. Tektronix, Inc., 1968.
17. S. M. Sze, Physics of Semiconductors Devices. Wiley, New York, 1967.
18. F. J. Hyde, Semiconductors. Macdonald, London, 1965.
19. R. L. Cummerow, Phys. Rev. 95,16 (1954).
20. J. C. Gammel and J. M. Ballantyne, Appl. Phys. Lett. 36, 149 (1980).
21. J. C. Gammel and G. M. Metze, IEEE Trans. Electron Devices ED-28,841 (1981).
22. C. Y. Chen, B. L. Kasper, and H. M. Cox, Appl. Phys. Lett. 44,1142 (1984).
23. A. Antreasyan, P. A. Garbinski, V. D. Mattera, Jr., N. A. Olsson, and H. Temkin, J. Appl. Phys. 60,1535 (1986).
24. R. A. Lawton, Electron. Lett., 11,74 (1075).
25. S. Y. Wang and D. M. Bloom, Electron. Lett. 19,554 (1983).
26. S. Y. Wang, D. M. Bloom, and D. M. Collins, Appl. Phys. Lett. 42,190 (1983).
27. D. H. Austin, Appl. Phys. Lett. 37,371 (1980).
28. C. J. Wei, H. J. Klein, and H. Beneking, Electron. Lett. 17,67 (1981).
29. R. C. Eden, Proc. IEEE 63,32 (1975).

30. R. D. Dupuis, J. C. Campbell, and J. R. Velebir, Electron. Lett. 22,48 (1986).
31. P. M. Downey, J. E. Bowers, C. A. Burrus, F. Mitschk, and L. F. Mollenauer, Appl. Phys. Lett. 49,430 (1986).
32. J. C. Campbell, A. G. Dentai, G. J. Qua, J. Long, and V. G. Riggs, Electron. Lett. 21,448 (1985).
33. J. E. Bowers, A. K. Srivastava, C. A. Burrus, J. C. DEwinter, M. A. Pollack, and J. L. Zyskind, Electron. Lett. 22,138 (1986).
34. S. R. Forrest, R. F. Leheny, R. E. Nahory, and M. A. Pollack, Appl. Phys. Lett. 37,322 (1980).
35. F. Capasso, R. A. Logan, A. Hutchinson, and D. D. Manchon, Electron. Lett. 16,893 (1980).
36. P. A. Wolff, Phys. Rev. 95,1415 (1954).
37. K. Seegar, Semiconductor Physics. Springer-Verlag, Berlin, 1982.
38. G. A. Barraff, Phys. Rev. 128,2507 (1962).
39. G. A. Barraff, Phys. Rev. 133,A26 (1964).
40. H. D. Law, K. Nakano, and L. R. Tomasetta, IEEE J. Quantum Electron. QE15,549 (1979).
41. F. Capasso, Surface Sci. 142,513 (1984).
42. R. D. Dupuis, J. R. Velember, J. C. Campbell, and G. J. Qua, Electron. Lett. 22,235 (1986).
43. D. Law, K. Nakano, and L. R. Tomasetta, IEEE J. Quantum Electron. QE15,549 (1979).
44. Y. Takanish and H. Horikoshi, Jpn. J. Appl. Phys. 17,2065 (1978).
45. H. D. Law, L. R. Tomasetta, K. Nakano, and J. S. Harris, "AlGaSb and AlGaAsSb High-Speed Avalanche Photodiodes," Device Research Conf. 1978, Santa Barbara, Cal.
46. H. D. Law, L. R. Thomasetta, K. Nakano, and J. S. Harris, Appl. Phys. Lett. 33,416 (1978).
47. H. Melchior and A. R. Hartman, Tech. Dig. Int. National Device Meeting, 1976, p. 412.
48. R. G. Smith, C. A. Brackett, and H. W. Reinbold, Bell Syst. Tech. J. 57,1809 (1978).
49. K. M. Van Vliet, Tech. Dig., IEDM, Washington, D.C., 1968, p. 298.
50. R. J. McIntyre, IEEE Trans. Electron Devices ED-19,703 (1972).
51. F. Capasso, IEEE Trans. Nucl. Sci. NS-30,424 (1983).
52. F. Capasso, W. T. Tsang, A. L. Hutchinson, and P. W. Foy, "The graded band-gap vavalanche diode," Int. Symp. GaAs and Related Compounds, Japan, Ser. 63,473 (1982).
53. F. Capasso, Surface Sci. 142,513 (1984).

54. F. Capasso, W. T. Tsang, and G. F. Williams, Trans. Electron Devices ED30,381 (1983).
55. F. Capasso, Ann. Rev. Mat. Sci. 16,263 (1986).
56. F. Capasso, K. Mohammed, and A. Y. Cho, IEEE J. Quantum Electron. QE22,1853 (1986).
57. F. Capasso, Science 235,172 (1987).
58. K. Kikuchi, T. Okoshi, and A. Hirose, Electron. Lett. 21,801 (1985).
59. J. R. Houck and D. A. Briotta, Jr., Infrared Phys. 22,213 (1982).
60. G. E. Stillman, C. M. Wolfe, A. G. Foyt, and W. T. Lindley, Appl. Phys. Lett. 24,8 (1974).
61. R. D. Dupuis, J. R. Velebir, J. C. Campbell, and G. J. Qua, Electron. Lett. 22,235 (1986).
62. Y. Matsushima, K. Sakai and Y. Noda, IEEE Electron Device Letters. ED1-2, 179 (1981).
63. D. L. Robinson and B. D. Metscher, Appl. Phys. Lett. 51, 1493 (1987).
64. R. Chin, N. Holonyak, Jr., G. E. Stillman, J. Y. Tang, and K. Hess, Electron. Lett. 16,467 (1980).
65. F. Capasso and W. T. Tsang, Electron Devices Meeting, San Francisco, 1982, p. 334.
66. N. R. Couch, D. G. Parker, M. J. Kelly, and T. M. Kerr, Electron. Lett. 22,636 (1986).
67. R. J. McIntyre, Trans. Electron Devices ED-13,164 (1966); ED-19,703 (1972).
68. F. Capasso, W. T. Tsang, A. L. Hutchinson, and G. F. Williams, Appl. Phys. Lett. 40,38 (1982).
69. V. D. Mattera, Jr., F. Capasso, J. Allam, A. L. Hutchinson, J. Dick, J. M. Brown, and A. Westphal, J. Appl. Phys. 60, 2609 (1986).
70. D. D. Coon, R. P. G. Karunasiri, and H. C. Liu, J. Appl. Phys. 60,2636 (1986).
71. H. L. Störmer, J. Phys. Soc. Jpn. Suppl. A49,1010 (1980).
72. R. Dingle, H. L. Stormer, A. C. Gossard, and W. Wiegmann, Appl. Phys. Lett. 33,665 (1978).
73. W. Jantsch, K. Lischka, A. Eisenbeiss, P. Pichler, H. Clemens, and G. Bauer, Appl. Phys. Lett. 50,1654 (1987).
74. G. Bauer, Superlattices Microstruct. 2,531 (1986).
75. H. L. Störmer, A. C. Gossard, W. Wiegmann, and K. Baldwin, Appl. Phys. Lett. 39,912 (1981).
76. E. E. Mendez, IEEE J. Quantum Electronics QE 22, 1720 (1986). This is a special issue on semiconductor quantum wells and superlattices. Several invited papers on photosensors represent present state of art of modern photodetectors.
77. T. Ando, A. B. Fowler, and F. Stern, Rev. Mod. Phys. 54, 437 (1982).

78. R. Fischer, D. ARnold, R. E. Thorne, T. J. Drummond, and H. Morkoc, Electron. Lett. 19,200 (1983).
79. Y. M. Pang, C. Y. Chen, and P. A. Garbinski, Electron. Lett. 19,716 (1983).
80. J. Bliss, Theory and Characteristics of Phototransistors, Motorola Semiconductor Application Note AN 440, 1974.
81. N. Chand, P. A. Houston, and R. N. Robson, Electron. Lett. 21,308 (1985).
82. U. Koren, T. C. Penna, and P. K. Tein, Electron. Lett. 21, 346 (1985).
83. F. Szmulowicz and T. H. Chandler, Jr., IEEE Trans. Electron Devices Ed28,772 (1981).
84. F. M. Klassen, K. M. van Vliet, and J. Blok, Physica 26,605 (1960).
85. D. Long and J. Schmidt, in Semiconductors and Semimetals, Vol. 5, R. K. Willardson and A. C. Beer, Eds., Academic, New York, 1970, p. 175.
86. P. Kruse, in Semiconductors and Semimetals, Vol. 5, R. K. Willardson and A. C. Beer, Eds., Academic, New York, 1970, p. 15.
87. D. Gasquet, H. Tijani, J. P. Nougier, and A. van der Ziel, in Noise in Physical Systems and 1/f Noise. M. Savelli, G. Lecoy, and J. P. Nougier, Eds., 1983, p. 165.
88. D. Gasquet, M. Fadal, and J. P. Nougier, in Noise in Physical Systems and 1/f Noise. M. Savelli, G. Lecoy, and J. P. Nougier, Eds., 1983, p. 169.
89. A. Antreasyan, C. Y. Chen, and P. A. Garbinski, Electron. Lett. 21,1137 (1985).
90. B. L. Kasper, J. C. Campbell, and A. G. Dental, Electron. Lett. 20,796 (1984).
91. F. Capasso, "Graded-gap superlattice devices by bandgap engineering" in Semiconductors and Semimetals, Vol. 24. R. K. Willardson and A. C. Beer, Eds., Academic, New York, 1970, p. 319.

7
Special Topics

The purpose of this chapter is to summarize, outline, and bring to notice a few approaches that are not frequently mentioned in the literature but could have significant importance in explaining photoconductivity phenomena in general.

7.1. EXTRINSIC PHOTOCONDUCTIVITY

Photoconductivity originates in the generation of mobile charge carriers caused by absorption of photons. In previous chapters we have considered that the valence electrons are ejected to the conduction band, giving rise to electron–hole pairs. It can also happen that electrons from impurity atoms are excited to the conduction band, leaving the ionized atoms behind. Excited electrons are mobile (since they are in the conduction band), and hence they will contribute to the conduction process. This type of photoconductivity is called extrinsic photoconductivity, because it is not an intrinsic property of the semiconductor but is caused by the presence of ionized impurities.

The energy required to ionize the impurity atom is of the order of a few millielectronvolts. The radiation corresponding to it lies in the far infrared (approximately 50–200 cm^{-1}), and hence extrinsic photoconductivity is observed in this range. Techniques of measurement are, obviously, different from those mentioned in Chapter 2. Because the ionization energy of the impurity atom is low, it ionizes at a temperature of about 10 K, and therefore the measurements have to be carried out at low temperature. Moreover, earlier experimental studies [1–5] showed that transitions from 1s to 2p or even to higher excited states are also observed in the photoconductivity spectrum. It is therefore customary to carry out these measurements at 4.2 K.

In the spectral range 10–500 cm^{-1}, dispersive spectrometers are not adequate, and Fourier transform spectrometers are generally employed. Moreover, because of recent developments in microprocessors, more sophisticated systems including finer control over movements of mechanical parts, fast and precise data acquisition, and improved software are available at relatively low cost, and therefore more and more laboratories are getting involved in the study of impurity photoconductivity.

The basic principle of Fourier transform spectroscopy is based on the fact that in a double-beam interferometer the intensity of the central fringe obtained at the output is the Fourier transform of the incident optical power spectrum [6,7]. The fundamental equation on which this method is based can be written as [6]

$$f(\nu) = \int_0^{\infty} \left[\underline{I}(x) - \frac{\underline{I}(0)}{2} \right] \cos 2\pi\nu x \, dx \tag{7.1}$$

where dx is the optical path difference introduced between the beams, $\underline{I}(0)$ and $\underline{I}(x)$ are the intensities at zero and x path differences, respectively, and $f(\nu)$ is the spectral intensity at frequency ν. In practice, intensities at the output of the interferometer are measured at discrete values of path difference, m dx, where dx is a step difference between two beams and m is the number of data points. In order to achieve a higher resolution, a greater number of data points (2^9–2^{12}, as the algorithm of the Fourier transform spectrum demands that the number of data points be expressed as a power of 2) must be recorded. Needless to say, it is given in digitized form so the Fourier transform can be calculated by numerical methods.

The other parameter that influences resolution is the source aperture, which has an inverse relation. The theoretical details of experimental information about improvements in resolving power have been discussed elsewhere [6], and hence are omitted here. A typical interferogram and its Fourier transform are shown in Fig. 7.1.

The theory, practice, and application of Fourier transform spectroscopy have been extensively reviewed in the literature (e.g., Ref. 8) and I will not repeat the details here. What is significant is that the Fourier transform spectrometer is a powerful tool for detecting radiation in the far-infrared region (40–200 cm^{-1}) with relative ease.

As we have seen in Chapter 4, the ionization energies of the donor or acceptor atoms and the separation between various excited

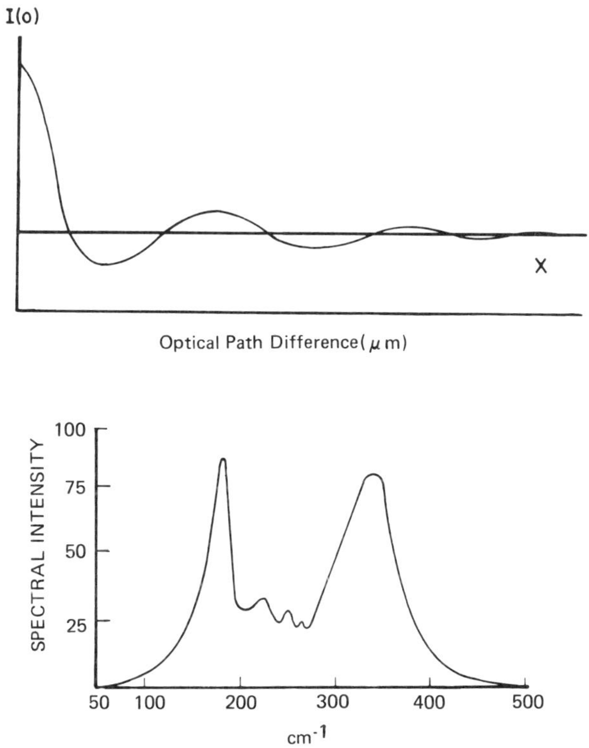

FIGURE 7.1 A typical interferogram and its Fourier transform, which shows the spectral distribution of radiation.

states (hydrogen-like bound energy states) can be estimated with the help of the equation [1]

$$E_n = -\frac{m^*e^4}{2\hbar^2\varepsilon^2n^2} \tag{7.2}$$

where ε is the dielectric constant, m^* is the effective mass of the electron, and n is a quantum number. Using known parameters (ε = 12.58 and m^* = 0.0665) for GaAs, the ionization energy can be calculated; it is found to be 5.72 meV (45.76 cm^{-1}). Thus, Fourier transform spectroscopy is the most suitable technique for investigating the energy states of the impurity atoms.

The obvious consequence of this gift in the hands of semiconductor physicists is the development of solid-state spectroscopy of shallow donors and acceptors (analogs of the hydrogen atom in solids) in high purity materials as the energy of ionization lies precisely in the region where the Fourier transform spectrometer is advantageous.

Surprisingly, experimental investigations show not only the transition corresponding to ionization of shallow donors in the optical absorption or photoconductivity spectrum but also the transitions corresponding to energy levels 1s to 2p, 3p, 4p (e.g., Refs. 9, 10), and rarely but clearly to the 5p and 6p levels [11,12]. Moreover, as mentioned earlier, the resolution of Fourier transform spectroscopy for higher values of data points (2^{11}) is good enough to resolve the separation of p energy states caused by external perturbation [12,13]. Naturally, external perturbations such as a magnetic field or uniaxial stress can be used to characterize the energy levels of the impurity atom and obtain more information about the site symmetry, local field effect, and so on. Zeeman spectroscopy of donors and acceptors is the next natural step, and in some high purity materials such studies have been carried out with remarkable success [14,15]. Further, the effect of uniaxial stress [16] has been examined and splitting of the energy states has been observed. On the basis of this information the site symmetry of donors or acceptors could be determined with reasonable precision.

Extensive research in elemental and compound semiconductors such as germanium, silicon, GaAs, and CdTe proves that binding energies vary with impurities of different atomic species. Tables 7.1 and 7.2 show the binding energies and ionization energies for various donors in germanium and silicon, respectively. The experimental data presented in these tables demonstrate the precision and resolution with which the spectrum of impurity atoms can be examined with this technique.

Both optical absorption and photoconductivity techniques have been employed in obtaining the above information. It is understandable that the energy corresponding to ionization should be observed in the photoconductivity spectrum, but it is not so obvious why the transitions from the ground state to the excited states (1s to np, n = 2, 3, 4, or 5) should appear in the photoconductivity spectrum, since they are localized. It is assumed that after optical excitation, thermal excitation [17] takes place, making available free electrons (in the case of donors) in the conduction band. The impact ionization process is also responsible in some materials, and considerable work has been reported in this area [18]. In some semiconductors the possibility of combination of photothermal and phonon-assisted processes has been examined by Seccombe and Korn [19] for a number of donors in germanium. Whatever may be the mechanism of ionization, it is very clear that photoconductivity spectra do show transitions from 1s to 2p, 3p, 4p, etc.

A large amount of information on shallow donor or acceptor states comes from extrinsic photoconductivity or optical absorption.

TABLE 7.1 Binding Energies of the Energy Levels of Donors in Germanium (meV)

Level	P[a]	As[b]	Sb[a]	Bi[b]
$1s(A_1)$	12.88,12.87[c]	14.18	10.45,10.32[c]	12.75
$1s(T_2)$	10.06[c]	9.94[c]	9.99[c]	9.90[c]
$2P_0$	4.72[c]	4.74	4.74[c]	4.75
$3P_0$	2.54[c]	2.57	2.57[c]	2.55
$2P_{\pm}$	1.72	1.74	1.71	1.67
$4P_0$				
$3P_{\pm}$	1.03	1.03	1.03	1.03
$5P_0$				
$4P_0$	0.75		0.75	
$4f_{\pm}$	0.62		0.59	
$5P_{\pm}$	0.47		0.46	
$5f_{\pm}$	0.39		0.39	
$6P_{\pm}$	0.32		0.32	
$6f_{\pm}$			0.30	

[a][Ref. 19.]
[b][Ref. 12.]
[c]R. A. Faulkner Phys. Rev. 184:713 (1969).

Both techniques are generally treated equivalently since they apparently give the same information. However, a closer look shows that in some materials the photoconductivity spectrum shows additional lines, which are interpreted as different chemical species of the impurity atom [20]. It seems, therefore, that the photoconductivity technique is more sensitive than optical absorption and can be used to detect lower concentrations of impurity atoms.

In fact, there is no genuine reason to assume that the photoconductivity technique is superior to absorption since the basic

TABLE 7.2 Binding Energies of the Energy Levels of Donors in Silicon (meV)

Level	P[a]	As[a]	Sb[b]	Bi[c]
$1s(A_1)$	45.59	53.76	42.74	70.98
1s(E)	32.58[b,d,e]	31.26[b,e]	30.47[b,e]	
$1s(E + T_2)$			32.89	32.89[f]
$1s(T_2)$	33.89	32.67[b]	32.91	31.89[f]
$2P_0$	11.48	11.50	11.51	11.44
2s		9.11[h]		8.78[h]
$2P_\pm$	6.40	6.40	6.38	6.37
$3P_0$	5.47	5.49	5.50[g]	5.48
3s	3.83[g]			4.70
$3d_0$	3.73	3.8[h]		3.80
$4P_0$	3.31	3.31	3.33[g]	3.30
$3P_\pm$	3.12	3.12	3.12	3.12
4s				2.89
$4f_0$	2.33			2.36
$4P_\pm, 5P_0$	2.19	2.19	2.20	2.18
$4f_\pm$	1.90	1.90	1.94[g]	1.91
$5f_0$	1.65		1.71[g]	1.67
$5P_\pm$	1.46	1.46	1.48[g]	1.46
$5P_\pm$	1.26			
$6P_\pm$	1.09	1.07[b]	1.10[g]	1.08

[a] [Ref. 5.]
[b] R. L. Aggarwal and A. K. Ramdas, Phys.Rev B,140A:1246 (1965).
[c] N. R. Butler, P. Fisher and A. K. Ramdas, Phys. Rev.B 12:3200 (1975).
[d] G. B. Wright and A. Mooradian, Phys. Rev. Lett. 18:608 (1967).
[e] K. Jain, S. Lai and M. V. Klein, Phys. Rev.B,13:5448 (1976).
[f] W. E. Krag, W. H. Kleiner and H. J. Zeiger, Proc. of 10th Conf. of Phys. of Semiconductors, Cambridge, Massachusetts. S. P. Keller, J. C. Hensel and F. Stern, Eds. U.S.A.E.C. Division of Technical Information, Washington D.C., p.271, 1970.
[g] B. Pajot, J. Kauppinen and R. Anttila, Solid State Comm. 31:759 (1979).
[h] W. H. Kleiner and W. E. Krag, Phys. Rev. Lett. 25:1490 (1970).

mechanisms for both processes are essentially the same. The fact that the photoconductivity spectrum shows more structure or weak peaks does not necessarily mean that it detects the presence of other species of the impurity atom. A contradictory opinion on the interpretation of extrinsic photoconductivity has been expressed by Joshi [21].

The presence of additional lines has been observed in semiconductors of silicon, germanium, GaAs, InP, and CdTe, where a substitutional atom has the same site symmetry (T_d) and p wave functions are not split up with T_d symmetry operations. However, when an electric field is applied, the elements of symmetry are reduced. Suppose that an electric field is applied in the ⟨100⟩ direction; then T_d symmetry becomes D_{2d} [22]. With the help of characters of D_{2d} symmetry, the splitting of a p wave function has been calculated, and it has been found that p wave functions are split into two components [21], which could give an additional line in the photoconductivity spectrum.

To test the validity of the above arguments, photoconductivity and optical absorption measurements were performed on high purity monocrystals of CdTe, and the additional structure in the photoconductivity spectrum was observed. Its intensity and separation increases with applied voltage [21] (see Fig. 7.2). A close comparison of photoconductivity and transmission spectra shows that peaks obtained with the former technique do not exactly coincide with the one obtained with the latter but lie on either side of it [20]. This

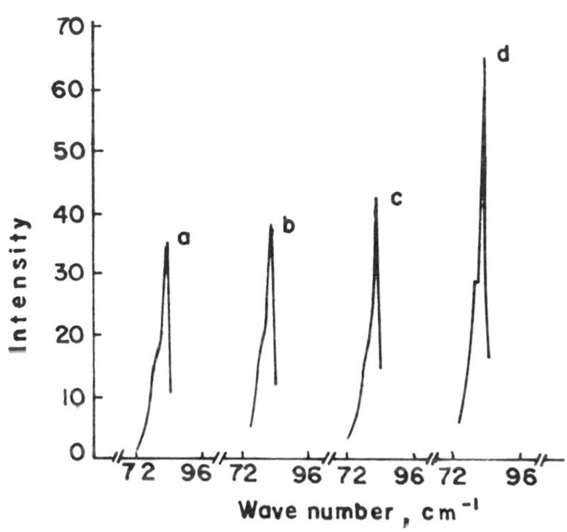

FIGURE 7.2 Electric field dependence of 1s → 2p transitions, reported in the photoconductivity spectrum of high-purity n-type CdTe (4.2 K). [After Ref. 21.]

supports an alternative approach proposed by Joshi [21]. Yet this explanation has not been accepted widely, because the magnitude of the splitting as a function of voltage has not been calculated precisely and direct experimental confirmation is still needed.

The above approach suggests that care should be taken to interpret the extrinsic photoconductivity data before concluding the presence of impurity atoms of a different chemical nature.

7.2. PHOTOIONIZATION OF DONORS OR ACCEPTORS

An electron from a donor atom can absorb a photon of proper energy and be ejected to the conduction band. This process is called photoionization of impurity atoms, and extrinsic photoconductivity can be used to examine it in detail. Theoretical calculations of the photoionization cross section depend on the approximations used in the initial wave functions, so reported expressions differ slightly from each other.

For simplicity, let us consider the ionization of the 1s electron of a shallow donor. In this case it is a good approximation to assume that the time dependent wave function of 1s electron of hydrogen atom is given by [23]

$$\psi_i = \frac{1}{\sqrt{\pi}} \chi^{-3/2} \exp(-\chi \cdot r) \exp\left(-\frac{iEt}{\hbar}\right) \tag{7.3}$$

where E is the energy of the 1s orbital. The significance of the term χ can be understood from the exponential term. At a distance χ^{-1}, the probability of finding the electron is reduced to e^{-1} and consequently can be considered a parameter that indicates the region of localization of the 1s electron of the impurity atom. Its magnitude depends on the impurity or defect concentration. This suggests that if the impurity concentration is increased, then χ will increase and the form of the photoionization cross section will be altered. Thus, the quality of the crystal should be connected to the form of the photoionization peak. A study in this direction was carried out by Joshi and Vincent [24], who compared the theoretical value of the cross section with the form of the experimentally observed absorption curve.

The photoionization cross section is given by [23]

$$\alpha_{\text{photoionization}} \propto \frac{\hbar\omega - E_I}{\hbar^2 \omega^2 [\hbar^2 \chi^2 + 2m^*(\hbar\omega - E_I)]^4} \tag{7.4}$$

where $\hbar\omega$ is the energy of the incident radiation and E_I is the ionization energy of the impurity atom. The value of χ depends on the purity of the sample under investigation. For high purity materials, χ has a lower value and therefore the absorption peak corresponding to the ionization energy is expected to be sharp. With this view, far-infrared spectra for very high purity monocrystals of CdTe and for a sample containing a relatively high concentration of impurities were obtained using the same experimental conditions of temperature, resolution, light source, and so on. These are shown in Fig. 7.3.

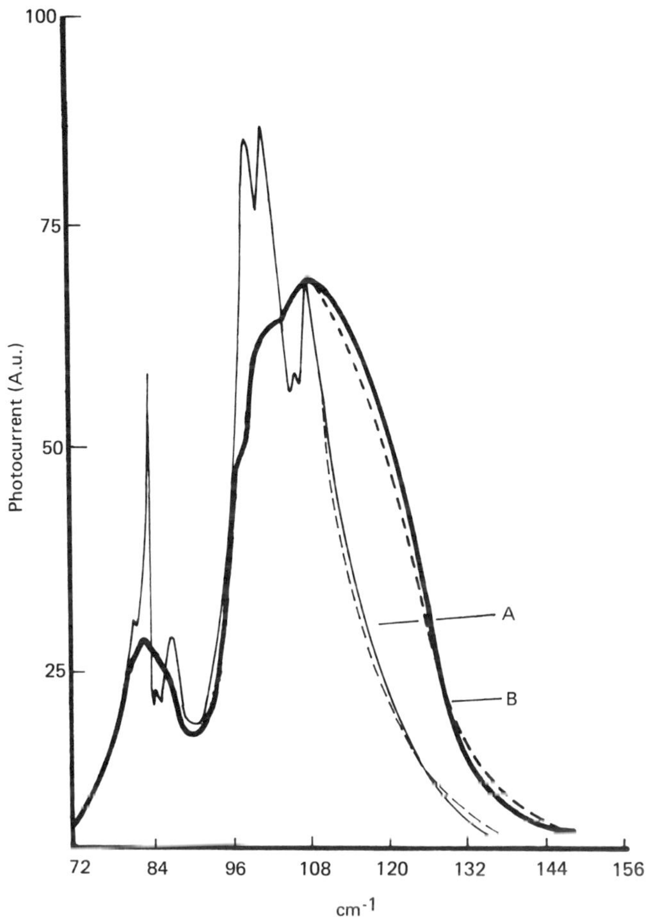

FIGURE 7.3 Photoconductivity spectra and theoretically simulated curves for photoionization cross section. The values of χ^{-1} are 210 Å for curve A and 160 Å for curve B, which correspond to highly pure and impurity introduced samples respectively. Dotted lines represent theoretically simulated curves. [After Ref. 10.]

Clearly, the peak corresponding to the ionization of the 1s level for a pure sample is sharp compared to the one obtained for a sample with an introduced impurity. This is a direct confirmation of the theory proposed earlier. As the extrinsic photoconductivity spectrum of CdTe has been studied extensively [25], the parameters used in Eq. (7.4) are known. Taking 12.3 meV as the ionization energy and $m^* = 0.096\ m$ [24,25], the photoionization cross section as a function of energy is simulated for various values of χ. The best fit is observed when χ^{-1} is equal to 210 Å for a high purity sample and 160 Å for a relatively pure sample. This is logical, since for the high purity sample the localization region for the 1s electron is higher than for the other sample. Further study and calibration of the broadening of the peak corresponding to the ionization process may be needed to evaluate the purity and quality of semiconducting materials. This potential capability could be a great asset in control of material quality for photonic devices.

Another interesting aspect of the extrinsic photoconductivity is the contribution from optical phonons [2]. The photoconductivity spectrum very often shows a series of equidistant minima originating from the emission of L.O. phonons. This is known as oscillatory photoconductivity, and considerable experimental work has been reported [26–28].

The position of the minimum in extrinsic photoconductivity can be given by [2]

$$h\nu = N_{phon}\hbar\,\omega_{phon} + E_I \tag{7.5}$$

where E_I is the binding energy of the impurity atom, N_{phon} is the number of phonons generated in the process, and ν is the phonon frequency at the center of the Brillouin zone. As expected from Eq. (7.5), the minima are found to be separated by Longitudinal Optical (LO) phonons close to the zone center and hence could be used to determine the energy of the phonons at the Γ point. The details of the process are given by Stradling [2].

7.3. PHOTOQUENCHING

"Photoquenching" is the term used to describe a phenomenon in which the photocurrent is reduced when additional radiation of different wavelength, sometimes called secondary radiation, is incident on the sample [29,30]. This behavior is generally attributed to the filling and emptying of certain defect states. The energy necessary to fill these states very often lies in the infrared region, and therefore the term "infrared photoquenching" is also used.

The probability of filling the defect states depends on their density, their positions with respect to the Fermi level, and their capture cross sections for electrons or holes. Naturally, it is not possible to predict in which crystal, with which type of defect states, and at what intensity and wavelength photoquenching will be produced, and hence the experimental study must be followed by adequate modeling to explain the observed behavior [31]. An exhaustive study of this property along with the temperature dependence has been carried out by Bube and Cardon [32] in monocrystals of CdS, CdSe, and GaAs, and some information about the positions of defect levels and their role in photoquenching phenomena has been determined. There are a few consistent models that necessarily deal with two types of defect levels in the band gap. Depending on the relative position of the Fermi level and light intensity, these levels behave like traps or recombination centers. These band-gap levels are always associated with different parameters such as capture cross sections and their densities and distribution. Because of the considerable variation in these parameters, a great deal of flexibility is created that facilitates the building of several models to explain not only optical or thermal photoquenching but also various related aspects such as sensitivity and superlinearity [33].

A typical model [33,34] generally used to explain photoquenching can be understood with the help of Fig. 7.4. For the given experimental conditions (i.e., for the intensity and energy of the incident radiation, primary and secondary) there exist two types of active centers. The first type are located close to the conduction band and have a high capture cross section for holes. The second type of centers are near the center of the band gap and have a higher capture cross section for electrons. By absorbing infrared radiation, electrons are excited to localized type II centers, and free

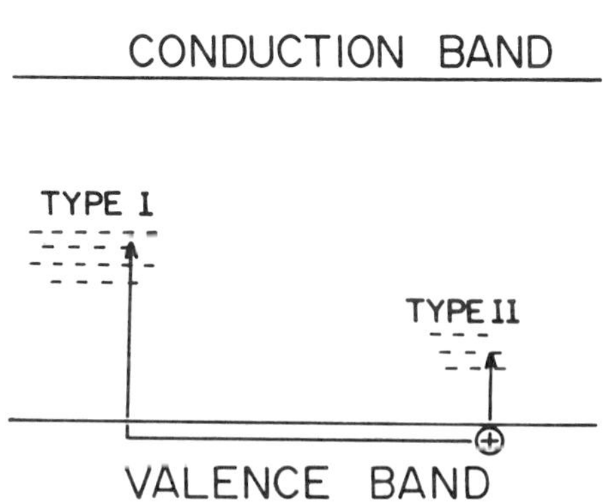

FIGURE 7.4 A two-level energy scheme to explain photoquenching.

holes are created that are immediately captured by type I levels. The result is that neither holes nor electrons participate in the conduction process. With this model, experimental data of photoquenching in several materials have been successfully explained.

The combined studies of various parameters of photoquenching such as temperature dependence, spectral response, and transient behavior along with thermally stimulated current in CdS and CdSe support this or other similar models. In order to explain the slight variation in the observed results, some minor changes are made by adjusting the density and capture cross section of the traps or locating the position of the Fermi level below or above the defect levels.

It is not always necessary to assume the presence of two types of traps or recombination centers. Optical quenching in iron-doped $KTaO_3$ has been reported by Yamaichi et al. [35]. Comparison of photoquenching and optical absorption spectra shows that the positions of photoquenching peaks show a similarity to those of the absorption bands observed in iron-doped crystals. This means that under secondary radiation, holes are excited from ionized iron sites and recombine with the electrons at type I centers and thus reduce the number of charge carriers. It seems that the iron atom generally accelerates the recombination process for one reason or another.

The above-mentioned mechanisms of photoquenching suggest that it is possible to determine with this technique one of the important parameters of traps, namely, their capture cross section and their variation with temperature. Such studies, although not common, have been carried out, and meaningful results have been reported [32]. A systematic study in this direction will be useful in characterizing the sample and avoiding these defects for photosensitive devices.

The majority of the proposed mechanisms necessarily require the presence of two sets of trapping/recombination centers, independent of their origins, within the band gap. Another model for photoquenching valid for p-type materials has been proposed by Joshi and Echeverria [36]. In this model, the number of electrons is not reduced by the trapping/recombination process through the energy states within the band gap, but the number of holes is reduced through recombination with the electrons ejected from the inner level of the valence band. A typical example is photoquenching in the ternary compound $CuGaTe_2$.

$CuGaTe_2$ is a p-type semiconductor with band gap 1.24 eV at 300 K [37,38]. Theoretical calculations show that the 3d level of copper lies 1.34 eV below the top of the valence band [39,40]. It is possible to excite electrons from these levels to the top of the valence band and reduce the number of holes that take part in the photoconduction process.

When the incident radiation has energy 1.24 eV (300 K), the absorption peak corresponding to the band-gap transition is observed; however, when the energy is slightly increased (1.34 eV), an unusual phenomenon occurs. Electrons from 3d levels of copper are ejected at the top of the valence band and therefore reduce the number of holes that are participating in the photoconduction process. Naturally, photoquenching is expected. Figure 7.5 shows the spectral response curve of $CuGaTe_2$ recorded at 300 K and 77 K. On the high-energy side of the meain peak a sudden dip due to photoquenching is observed, as expected [36]. Recent continued endeavours in this direction show that this is not an isolated example. The sudden steep reduction in the photoresponse on the high energy side of the band gap has been also observed in iron doped InP [41]. In this case, inner Fe^{++} level is located 1.5 eV below the top of the valence band where as band gap is 1.45 eV at 77 K.

When the magnitude of the separation between the top of the valence band and the deep level located in it is on the order of the band-gap energy, then the photoconductivity spectrum should have an unpredicted form as the holes created by the band-gap transitions are partially eliminated by the electrons ejected from the deep-level transitions. Thus, the consequence of photoquenching, in

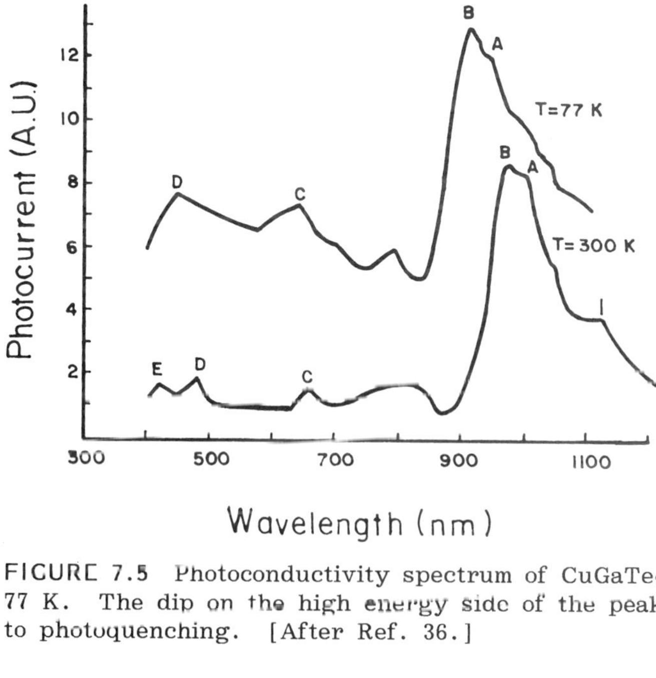

FIGURE 7.5 Photoconductivity spectrum of $CuGaTe_2$ at 300 K and 77 K. The dip on the high energy side of the peak is attributed to photoquenching. [After Ref. 36.]

this case, is to alter the form of the photoconductivity spectrum, and this is done to such an extent that the observed spectrum is difficult to interpret. Consideration of photoquenching due to inner-level transitions helps to clear the understanding of the photoconductivity spectra of both zinc- and silicon-doped GaAs (p type) [41,42].

Optical absorption and photoconductivity spectra of zinc-doped GaAs were investigated at 300 K; the results are shown in Fig. 7.6. In this case, there is no one-to-one correspondence between the two spectra; the photoconductivity peak lies on the lower-energy side of the absorption maximum and also on the lower-energy side of the accepted value of the forbidden gap. A close look at these spectra shows that there is a weak but noticeable peak on the lower-energy side of the absorption edge.

An apparently unexplainable aspect of the present result is that the minimum in photoconductivity is observed where the absorption maximum is located. This can only be explained by the photoquenching phenomena caused by the ejection of electrons from the deep level situated in the valence band at 1.37 eV (energy is measured with respect to the top of the valence band). In this circumstance, the photoconductivity peak observed at 1.28 eV is not a peak in the true sense (does not represent a band-gap transition), but photoquenching on the higher-energy side of 1.28 eV makes it appear to be a peak.

Even though the photoquenching model proposed here explains the results satisfactorily, the origin of the level located at 1.37 eV is not clear yet. It is known that such states are created in the valence band due to the absorption of oxygen and/or other gases. Depending upon the quantity of the gas, the cleavage plane, and other surface conditions, the Fermi level may be pinned or remain free [44]. If it is pinned, sharp electron distribution curves are observed below the valence band. Our theoretical knowledge of surface interaction is not developed enough to enable calculation of the precise position and electron distribution curves, and therefore the reason for the deep valence level is unclear. The consequence of this approach is that it helps to understand the previously observed but never explained form of the photoconductivity spectrum (such situations are not frequent; see, e.g., Ref. 43).

Of course, the shifting of the photoconductivity peak on the low-energy side is frequently explained on the basis of the energy band tail states caused by the spatial fluctuations in potential originating from the impurity atoms. However, this explanation does not help to understand the apparent anomaly in the comparative study of absorption and photoconductivity spectra.

Reduction in the photocurrent is observed not only with excitation of secondary radiation or ejection of electrons from deep levels

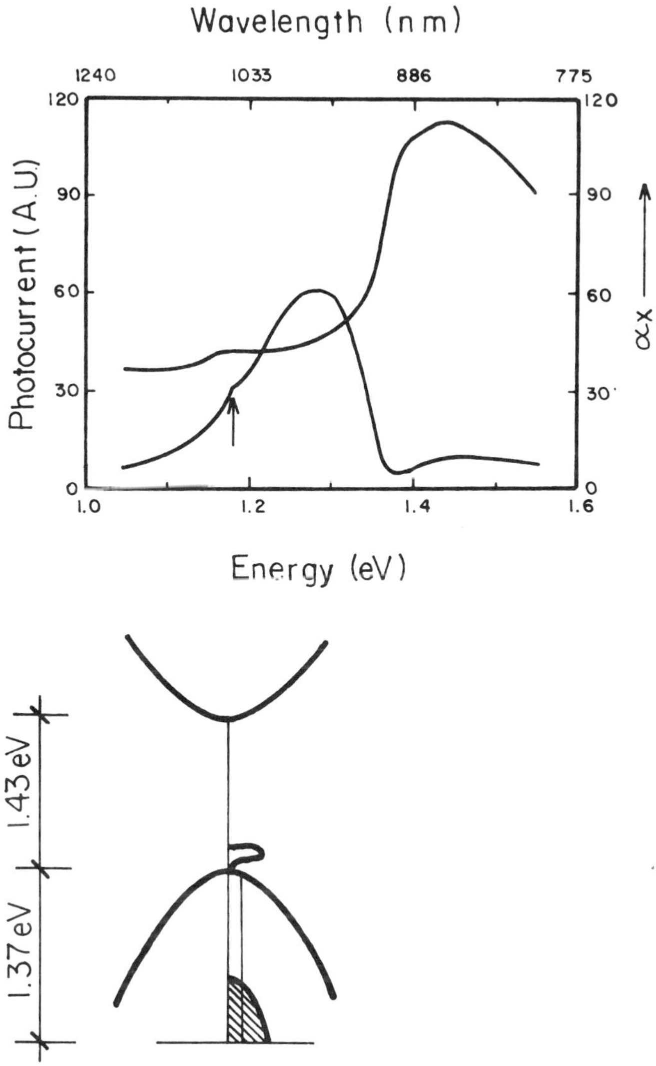

FIGURE 7.6 Photoconductivity and absorption spectra of p-type GaAs with the corresponding energy scheme. [After Ref. 42.]

but also upon application of an electric field [43]. This is called "field-induced photoquenching." This behavior is not very common but is reported in the literature from time to time, particularly in polycrystalline or inhomogeneous photoconductors [44]. In this case, it is observed that for a certain range of applied voltage, photocurrent decreases even though voltage increases, and hence this phenomenon is also referred to as negative differential photoconductance.

Models to explain the above effect are generally based on the assumption that there exists a region of potential barrier that separates slow and fast recombination centers from each other. When radiation of energy equal to that of the band gap is incident on the sample, then holes are captured by the slow recombination centers. For a certain value of voltage, and in the presence of radiation, the potential barrier is reduced, permitting the flow of holes from slow recombination centers to fast ones, and hence the photoresponse is quenched.

There are several possible causes for the potential barrier (e.g., inhomogeneity, grain boundary, and so on) and for slow as well as fast recombination centers, and the actual causes are determined on the basis of the parameters of the traps present in the particular photoconductor under discussion. Here we have dealt only the general approach and not the details.

7.4. NEGATIVE PHOTOCONDUCTIVITY

In a few cases it has been observed that when radiation of a certain energy is incident on a photoconductor, a decrease in current is observed instead of the expected increase; this phenomenon is called "negative photoconductivity." The variation in the mobilities of charge carriers is generally very small (here two-dimensional materials are excluded) and hence can be ignored. This means that either the number of free charge carriers or their lifetime is substantially reduced by incident radiation. Experimental observations of negative photoconductivity are contrary to normal expectations, but, occasionally, reported data confirm such behavior.

Negative photoconductivity has appeared in the literature occasionally ever since 1877 [47], initially in bulk materials such as germanium [45], silicon [46], selenium [47,48], cadmium sulfide [48], and cadmium selenide [50], and, later, in thin films [51]. Even though in the initial stages it was difficult to understand the origin for this behavior, now the process is intelligible. One well-accepted model due to Stöckmann [52] is based on a two-level scheme. According to this model, the forbidden band gap contains two types of centers with energies E_1 and E_2 as shown in Fig. 7.7. One type is located between the Fermi level and the conduction band, while the other is situated close to the valence band or between the Fermi level and the valence band. It is also assumed that the first type of centers have a high capture cross section for electrons and the probability of electrons being ejected to the conduction band is very

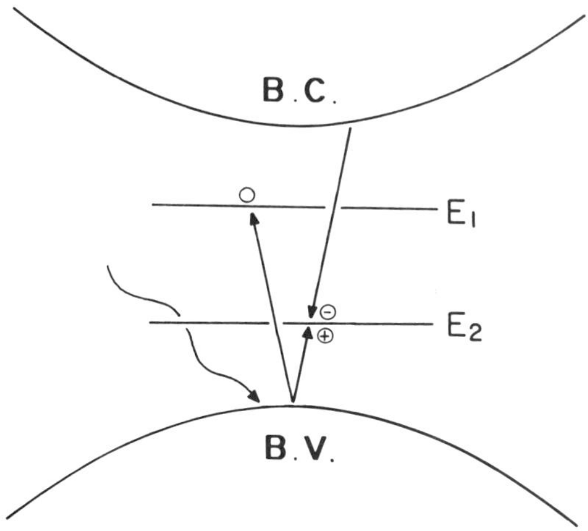

FIGURE 7.7 Energy scheme to explain negative photoconductivity (proposed by Stockmann [52]).

low. In short, the function of these types of centers in the presence of radiation is to create holes (by accepting electrons from the valence band) but at the same time not to increase the number of free electrons.

The second type of centers have a high cross section for electrons and holes, and consequently they capture electrons from the conduction band and holes from the valence band and recombine them. Thus, the net number of mobile charge carriers is reduced due to incident radiation, giving rise to negative photoconductivity.

Other aspects of negative photoconductivity, such as its temperature dependence and its variation in magnitude with intensity of illumination and with bias voltage, have been examined by Abdullaev et al. [53] in the ternary compound $TlGaSe_2$. In this case, a three-level scheme was needed. I wish to stress that the simple model proposed by Stöckmann explains directly or with slight modification almost all experimental observations of negative photoconductivity.

Very recently an alternative model to explain the negative photoconductivity in p-type semiconductors has been advanced [54]. The approach is similar to the one taken for explaining photoquenching from the inner levels. This model essentially differs from the model proposed by Stöckmann, in the sense that there is no need to assume a two-level system in the forbidden gap; it is proposed that electrons ejected from the inner level can recombine with holes at the top of the valence band and bring about negative photoconductivity. This phenomenon has been demonstrated in $Cd_{1-x}Fe_xSe$ [54].

Figure 7.8 shows the photoconductivity spectrum of $Cd_{1-x}Fe_xSe$ recorded at 300 K. The absorption spectrum is also shown for comparison. Like photoquenching, there is no one-to-one correspondence between them. At 1.7 eV, photocurrent shows a sudden drop where the absorption continues to increase. The reduction in the current was so great that the current (in the presence of radiation) is lower than the dark current, which was confirmed by the oscillogram shown in Fig. 7.9. After irradiation, the photocurrent increases, but after 2 sec the photocurrent starts to decrease and reaches its negative value.

As there is no evidence to suggest that there are two types of traps in the forbidden gap, negative photoconductivity is explained on the basis of the known energy-level scheme of iron in $Cd_{1-x}Fe_xSe$, which is shown in Fig. 7.9b [54]. It is clear that radiation of energy 1.72 eV excites electrons from the inner level of iron to the top of the valence band, where they recombine with the holes and the photoconductance attains a value lower than dark conductance.

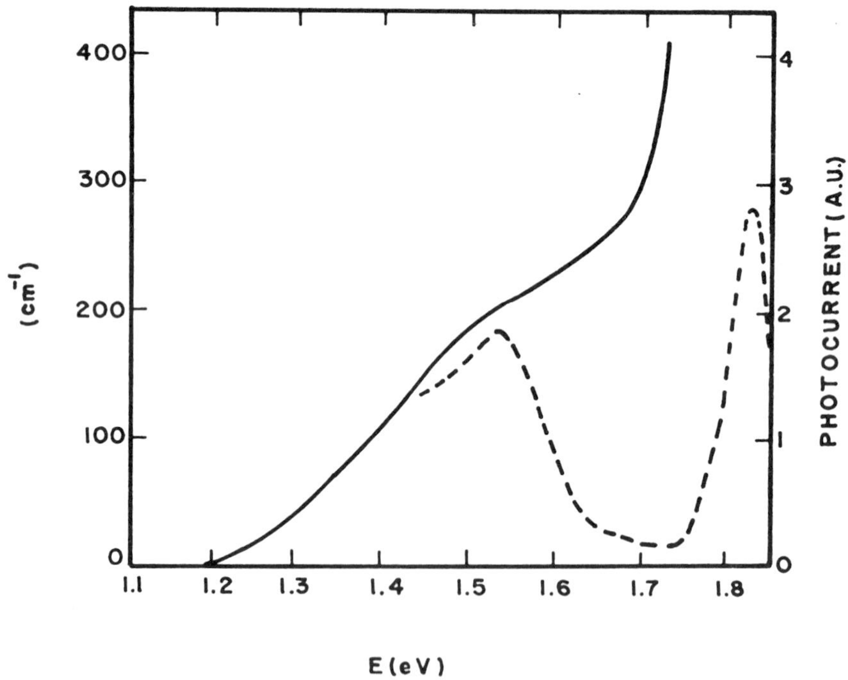

FIGURE 7.8 Photoconductivity and optical absorption spectrum of $Cd_{1-x}Fe_xSe$. [After Ref. 54.]

(a)

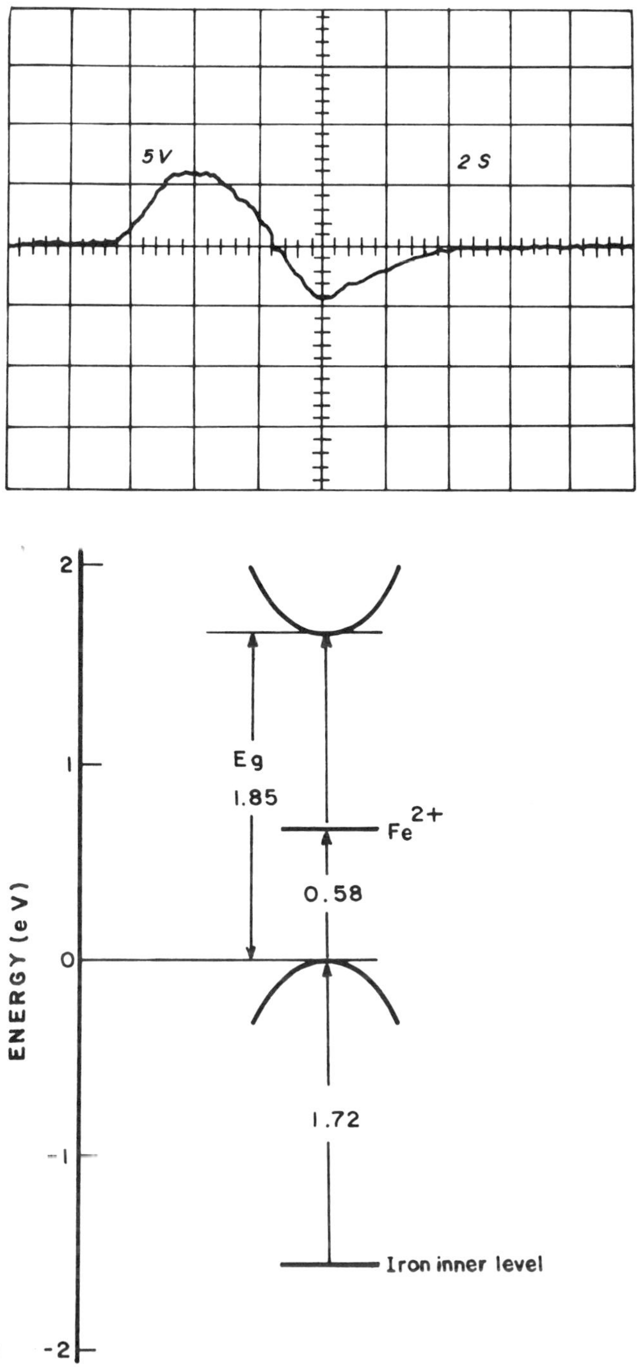

(b)

FIGURE 7.9 (a) Oscillogram indicating negative photoconductivity. (b) Energy-level diagram to explain the observed negative photoconductivity [54].

Negative photoconductivity is particularly expected in p-type diluted magnetic semiconductors, where the role of 3d levels of the transition atoms is dominant. There are two special features of this model.

1. It explains the reduction in current due to monochromatic radiation.
2. It is applicable to low-band-gap materials where two or more types of localized levels are generally not observed.

Very recently, extremely high negative photoconductivity was reported by Höpfel [55] in a p-type modulation doped GaAs-AlGaAs quantum well structure at low temperature. Contrary to the observation mentioned earlier, in this case the photoresponse is negative over the entire spectral range. This suggests that negative photoconductivity in quantum well structures has different types of origins.

High negative photoconductivity for the entire spectral range cannot be due to a single mechanism. According to Höpfel, photoexcited charge carriers are separated by the strong internal electric field originating from ionized acceptors and the hole plasma. Holes are collected in the potential minima of the high-band-gap material, whereas electrons are collected in the quantum wells. Electrons and holes are recombined rapidly, reducing the total concentration of the mobile charges, and hence create a negative photoconductance. This tentative explanation is valid only for the high-energy region (600—700 nm) in which photon energy is sufficient to create pairs of charge carriers.

In the lower-energy region of the spectrum (700—800 nm), the negative photoconductivity is explained by other mechanisms as the photon energy is too low for excitation of holes into higher-band-gap material (e.g., AlGaAs). Here it is assumed that there exist donor-like traps that are ionized by incident radiation, and the extra electrons that are generated help to increase the recombination process by the presence of plasma. Thus, the number of holes is reduced substantially. The other suggested mechanism is a substantial contribution from the electron—hole scattering process. This requires that the carrier lifetime be longer than or approximately equal to a few nanoseconds. This demand is not always fulfilled for the GaAs-AlGaAs structure, and hence a separate investigation in this direction is needed.

In two-dimensional solids there are several additional complications such as the possibility of absolute negative mobility, spatial variation of band structure, and confinement of charge carriers and its effect on the rate of recombination. Our present knowledge of these aspects does not permit a complete understanding of negative photoconductivity in two-dimensional structures.

7.5. PHOTOSENSITIVITY

This topic has been discussed extensively since 1960 (see, e.g., Ref. 29) because of its fundamental importance in photodetection technology. Several highly sensitive commercially available detectors such as CdS, CdSe, and CdTe are not very sensitive in their intrinsic form but can be made sensitive by introducing a proper type of impurity or impurities: donor—acceptor or both [29,55]. This means that additional centers, traps and/or recombination sites, are created in the crystal. The immediate consequence of the presence of traps is to increase the lifetime of the charge carriers, and therefore these detectors have a high response and consequently slow speed—of the order of milliseconds to seconds.

Parameters that influence photoresponse can be understood easily with the help of the equation

$$I_{pho} = g_0 e \mu_n \tau_n \tag{7.6}$$

Here we neglect the contribution from holes.

Just a few years ago it was thought that a substantial change in the value of μ_n was rather unlikely even though the quality of the material was improved. However, recent developments in new materials such as superlattice structures, interfaces at heterostructures, and modulation doped materials have shown that mobility could be increased by a few orders of magnitude. Thus, not only is the magnitude of the response increased but also the response time is reduced, and modern photodetectors enjoy both advantages.

In conventional three-dimensional materials, the major increase in photoconductivity can be achieved only by increasing the number of free electrons and their lifetimes simultaneously. These requirements are fulfilled by the incorporation of proper impurities.

In the above discussion, we emphasize the term "proper impurities" because there are some types of impurities that instead of increasing the response, reduce it, and there are others that are not sensitive at all. Moreover, some impurities are sensitive within certain temperature ranges and at certain levels of light intensity. The role of the impurities, obviously, is not straightforward, and it is not possible to generalize. Every situation must be evaluated separately.

Much of the previous work on photosensitivity refers to the incorporation of impurities in II-VI compounds [29,56—58], and therefore we focus our attention on these compounds, but similar arguments can be extended to other materials.

If an atom of group II or VI of a II-VI compound is substituted by an atom of group III or VII, respectively, then there will be an extra electron at the cation or at the anion site, that is, the impurity atom of these elements will act as a donor. Similarly, if they are replaced by atoms of group I and group V, then there is a lack of electrons and an acceptor state is created. It is worth pointing out that the vacancies at anion sites are equivalent to donors since the electrons from cations are not used in the binding but they are free. In the same way, vacancies at the cation sites act as acceptors. The ionization energies for several donor and acceptor states for II-VI compounds are available in the literature and are summarized by Bube [29]. Similar information for other materials is also available.

The importance of the ionization energies, and relative positions of donor/acceptor states with respect to the quasi-Fermi level for electrons and holes, can be understood with the role of these states in the sensitization process. The model given below is widely accepted and is often used in discussing photosensitivity. Here only the basic principle is outlined. The details of the model are described by Rose [33] and Bube [29].

Let us consider the energy-level diagram shown in Fig. 7.10a. If the defect levels are situated above the quasi-Fermi level for electrons and below the quasi-Fermi level for holes, then these states are virtually inactive and sensitivity is not affected. If the defect states are such that they capture the free electrons for a long time and recombine by any process (radiative or nonradiative), then the response will be decreased. Sensitivity is increased generally by introducing acceptors, for example, copper in CdSe. The energy states created by these types of centers are referred to as type II and have a higher capture cross section for holes than for electrons (opposite to type I centers). The holes are captured by these states and remain longer than that of the type I states. It has been shown [29,56] that in the presence of radiation there is a tendency of charge transfer from type II state to type I state. As a higher number of centers of type I contain electrons, the probability of capturing them from the conduction band is reduced.

In short, conduction electrons more frequently see the centers with lower capture cross section (type II) than the centers with higher capture cross section (type I, since they are occupied), and this increases the lifetime of the electrons and therefore increases the photoresponse. The activities of these centers depend on their relative positions with respect to the quasi-Fermi level; the latter shifts with the temperature and the intensity of the radiation, and therefore the sensitivity varies both with temperature and intensity of radiation.

Here we describe a general trend of arguments used in this or other similar models. The details of the mechanism may differ; what is common in these models is that the set of localized states in the band gap increases the lifetime of the carriers, and according to Eq.

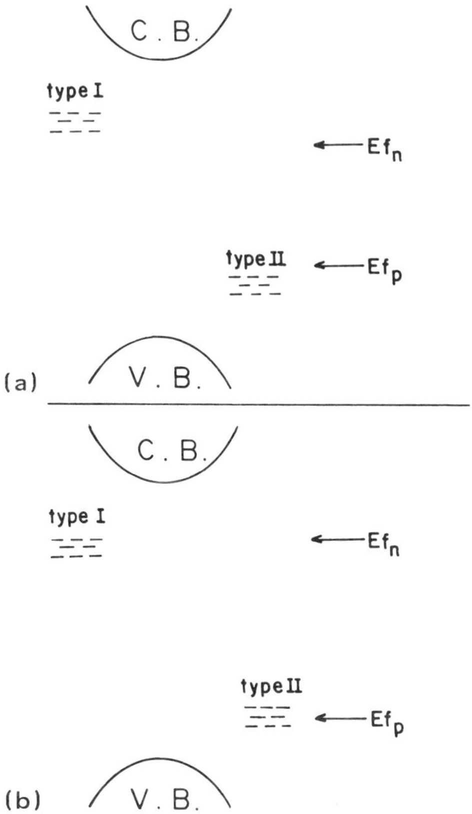

FIGURE 7.10 Commonly used energy scheme of defect states. (a) Low intensity of radiation; (b) moderately high intensity of radiation.

(7.6) the sensitivity of the photoconductor increases substantially (by as much as six orders of magnitudes).

In the literature, a few methods are cited for increasing the sensitivity of a given material, but they are essentially the same except that the method of creating trapping or recombination centers is different. For example, traps can be created by irradiating the sample with fast electrons [57]. Depending upon the density and distribution of the traps, photosensitivity can be substantially enhanced. This process is known as sensitivity due to irradiation and has been reported several times. A typical example is photoenhancement in CdS [58]. It has been found that 10-MeV electron irradiation increases the response and the major contribution comes from the peak located on the low-energy side of the band gap. This confirms that the radiation-produced defect states are responsible for the observed photosensitivity. In such circumstances, the obvious method for locating the position of optically active centers is to employ photoconductivity studies.

The other commonly used method for creating traps is to anneal a sample at its critical temperature, and recently this was used to increase the sensitivity of lead monoxide considerably [59]. Annealing seems to increase photosensitivity in several II-VI compounds [60].

The technique of material growth also plays a crucial role in determining the sensitivity of a given detector. For example, Si-Ge alloy grown by glow discharge plasma shows a high photoresponse [61]. In fact, the different growth methods and/or annealing create and stabilize the defect centers, which is the central part of photosensitivity, in different ways. More important than the method of creating a set of traps is their relative positions with respect to the Fermi level and their capture cross section for electrons and holes. These parameters determine whether the sensitivity will increase, decrease, or remain unaltered.

Generally, the material in which sensitivity is increased by any of the above-mentioned processes shows a photoresponse peak slightly on the low-energy side of the forbidden gap, since a substantial contribution comes from the transitions associated with donor or acceptor states. However, in some ternary semiconductors, such as $CdIn_2S_4$, the sensitivity is noticeably increased on the higher-energy side of the forbidden gap [62,63]. To understand the reason for this behavior a totally different outlook is necessary.

Like many other semiconductors, $CdIn_2S_4$, in a stoichiometric form, is a poor photoconductor but shows a remarkable response when enriched with sulfur. To examine the reason for this a systematic study was carried out and the spectral response for successive stages of sulfur enrichment was recorded. Figure 7.11 shows the spectral response for $CdIn_2S_4$ nominally enriched with sulfur and highly enriched with sulfur. In the latter case, a high-intensity peak is observed on the high-energy side and a broad peak on the low-energy side. The latter is expected because of the presence of impurity states within the band gap. The photoresponse for energy higher than the band gap is attributed to the high density of energy states created in the perturbed valence band structure due to the excess of sulfur atoms. Thus, there are some situations in which high sensitivity stems from a high density of states in the valence band rather than from defect states located within the forbidden gap. In addition to nonstoichiometric composition, some adsorbed gases can also create such a situation.

It has long been known, particularly for II-VI compounds, that a remarkable change is observed in the photosensitivity of the com-

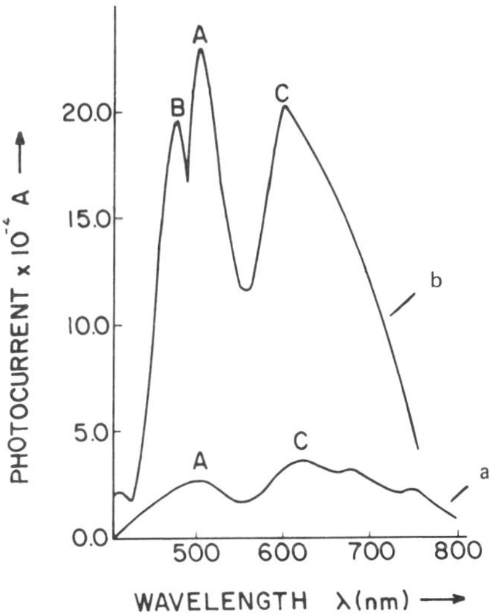

FIGURE 7.11 Photoconductivity spectrum of $CdIn_2S_4$. (a) Stoichiometric; (b) excess of sulfur. [After Ref. 63.]

pound when gas particles are adsorbed on the semiconductor surfaces. In the case of oxygen, the sensitivity is increased, [29] while in the case of water vapor a reduction in sensitivity is observed. The oxygen sensitization process, as usual, is attributed to the creation of oxygen acceptor states, and a few models have been proposed to explain the observed behavior [29]. Without going into the details of the mechanisms, the rise in the photoconductivity is explained by assuming that the adsorption-induced band-gap states capture electrons or holes and thus reduce the direct rate of recombination. This increases the lifetime of the charge carriers and hence the photosensitivity.

It has been found experimentally that if the gas is pumped off, the original state of photoconductivity is achieved, that is, the observed effect is reversible. This suggests that chemical reaction does not take place at the surface layers and gas atoms or molecules are not diffuse in the bulk material but are attached to the surface. Considering that adsorption-induced states are created within the band gap, the models are developed to explain the sensitization centers. These widely accepted models need further reexamination as photosensitivity is a key parameter in photodetection device technology.

Recently, this aspect was investigated thoroughly, both theoretically and experimentally, for understanding the origin of the oxygen sensitization process [64]. For this purpose, an n-type CdTe single crystal was used, because on the basis of ultraviolet photoemission spectroscopy and electron loss spectroscopy it has been strongly suggested that adsorption states of O_2 do not exist in the band gap [64–67]—a key parameter of the proposed model. This could be a decisive experiment. Toward this end, photosensitivity measurements were carried out for a few temperatures and for a few doses of O_2 and N_2.

Figure 7.12 shows photoconductivity spectra recorded at 77 K for two doses of O_2 [64]. It is clearly seen from the two spectra that there is a noticeable increase in photosensitivity as the adsorption increases. For the higher dose, the photoconductivity peak is intense and broad. Keeping the same experimental conditions, the dose was reduced. The intensity of the peak was also reduced and a new structure (peak A in Fig. 7.12) was revealed. By varying the doses of oxygen, it was confirmed that the intensity of the peak B, located on the high-energy side of the band gap, increases with adsorbed oxygen.

These experimental findings directly suggest that the high response is due to transitions from the energy states located below the top of the valence band and not within the band gap, as suggested traditionally [29]. The positions of these states should depend on the nature of the adsorbed gas and therefore should vary with the type of gas. This was confirmed by examining the effect of adsorption of nitrogen on the photoconductivity spectra. The observed results were similar to those just discussed except that the position of the peak located on the higher energy side was shifted. This suggests that surface adsorption induces high-density states in the valence band that are responsible for high photosensitivity.

An additional confirmation for this theory has been obtained by examining the temperature dependence of adsorption-induced peaks (see Fig. 7.13). As the temperature increases, desorption takes place, that is, the quantity of adsorbed gas is reduced, resulting in a reduction in the density of states in the valence band and therefore in the photoresponse. This can be seen through the reduction in the relative intensity of peak B with respect to peak A (corresponding to a band-gap transition). Experimental investigation [63] also shows that the same is true for adsorption of N_2 gas, the only difference being in the position of the density of states in the valence band.

In the last few years extensive investigations, both theoretical and experimental, have been carried out on surface states in semiconductors. There is evidence [65–67] obtained from photoemission studies, low-energy electron diffraction photoemission, and ultraviolet spectroscopy that in semiconductors, specifically in II-VI and III-V compounds, the adsorption-induced filled states are created below the top of the valence band. This conclusion is also confirmed by theoretical studies. The precise position of the high

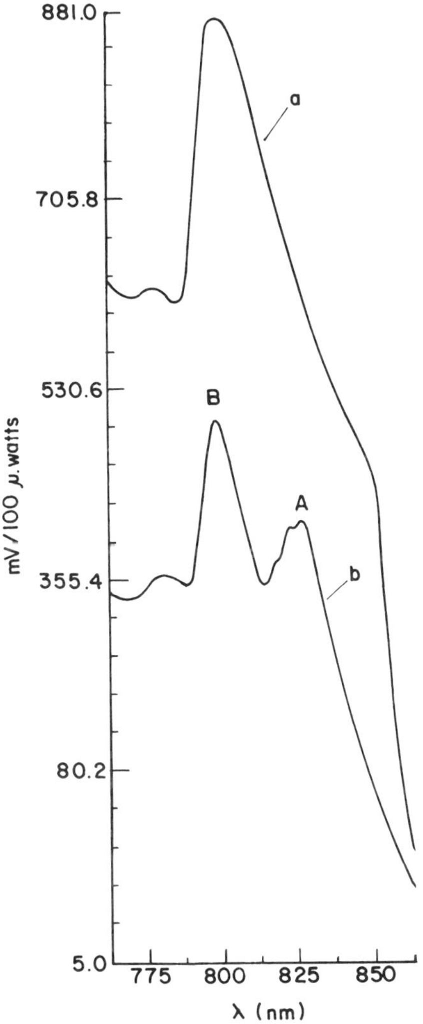

FIGURE 7.12 Adsorbed induced photoconductivity recorded at 100 K for different doses of O_2. (a) Adsorbed dose 5×10^6 L; (b) adsorbed dose 5×10^3 L. [After Ref. 64.]

density of states depends on the nature of the adsorbed gas, the ionicity of the material, and the specific plane on which the adsorbed particles are situated, among other things. Considering the complexity of the existing models and the imprecise role of the parameters, it is not possible to determine the position of the level below

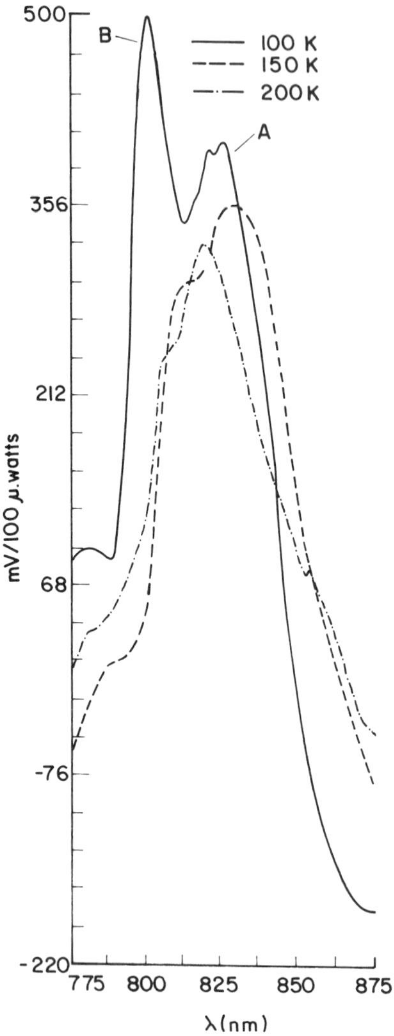

FIGURE 7.13 Temperature dependence of peak B due to oxygen adsorbed states in the valence band. [After Ref. 64.]

the top of the valence band. What is clear is that the adsorbed particles create extra energy states in the band gap, resulting in high photosensitivity in many II-VI and III-V compounds.

The revised interpretation presented here has two important consequences. First, the photodetection technique could be used to detect the adsorbed gases, and second it can provide information about the energy levels located below the top of the valence band where sophisticated techniques such as photoemission and low-energy electron spectroscopy are generally employed.

7.6. PERSISTENT PHOTOCONDUCTIVITY (PPC)

A curious and one of the least known photoelectric effects in semiconductors is photoconductivity that persists for a long time (minutes, hours, days, or even weeks) after the optical excitation that caused it is removed. It is also a puzzling effect, since the relation between the radiation that was once incident on the sample and the charge carrier generation or recombination rate is not obvious. Because of the potential practical applications in this field, particularly in optoelectronic memory devices, considerable work has been done [68—72] in the last few years and a significant information is available now. In spite of reasonable data, quantitative theory for PPC has not yet been developed.

One of the general features of persistent photoconductivity is that it can be quenched by heating the sample above some characteristic temperature, which depends on the material and the type of impurity present in it. Many semiconductors exhibit this behavior at low temperature, at about 77 K [70], but it is not uncommon to observe it even at room temperature.

This phenomenon has not been limited to any particular class of semiconductors but has been encountered in a great variety of materials. The list includes conventional II-VI, III-V, and ternary ($CdIn_2S_4$, $ZnIn_2S_4$) compounds as well as amorphous semiconductors and organic dyes. Persistent photoconductivity has also been recorded in materials in a variety of forms: thin homogeneous and polycrystalline films, single crystals, compacted pellets, and even powders. The slow component can also be induced by irradiating the sample with high-energy radiation such as X-rays or electrons (i.e., creating different types of defects). Considering the varieties of ways for obtaining PPC, it is rather difficult to grasp the mechanism coherently.

A few models exist to explain the remarkably slow and persistent photoresponse. Conventionally, these models are based on one of the following peculiarities of photoconductors.

1. Microscopic or macroscopic potential barriers separate charge carriers in real space and reduce the probability of recombination between them.
2. An energy state in the forbidden gap stores the charge carriers for a considerable amount of time before ejecting them to the conduction or to the valence band.

3. Photochemical reactions [70], may create new trapping and recombination centers along with easily ionizable shallow donor levels. In this case PPC is enhanced due to optically created donors and decays with the time constant corresponding to that of donors.
4. The energy of the incident radiation is enough to deform the photosensitive material and perturb the localized energy states near the deformation sites.

One of the above features is used to develop the model for PPC.

The duration of time for which persistent photoconductivity is observed depends upon the details of the origin of the potential barrier or types of traps or defect centers, their capture cross section for electrons and holes, the probability of ejection of charge carriers to the conduction or valence band, and other factors [74–76].

Macroscopic potential barriers such as junctions and surface barriers are very common in thin films, particularly in those cases where the layers of different carrier concentrations are grown one over the other. The model is successfully used to explain PPC in vapor-phase epitaxial GaAs (n-type) grown on chromium-doped semi-insulating GaAs substrate. A potential barrier similar to an n-p junction is created due to the relative difference between Fermi levels. The mechanism for the junction relaxation process, which is responsible for the delay component of PPC, can be visualized by considering the p-n junction under illumination. As we have seen in Chapter 6, the radiation reduces the height of the barrier, and the forward current cancels the reverse current originating from charge separation. When the radiation is turned off, the forward current will continue to flow until the barrier becomes sufficiently high to stop the flow. The time required for this depends upon many factors such as the densities of charge carriers on both sides of the junction, width of the depletion layer, and dielectric constant of the medium. The calculation and details of this model are given by Farmer and Locker [73].

We have seen in Chapter 6 that because of the ability to control the growth and processing technology of AlGaAs-GaAs heterostructures, several photoconducting devices are based on this system. The number of defects at the interface in these devices is very small, and in spite of that, in the majority of cases PPC has been reported (see, e.g., Ref. 78). Even though the origin for the slow component is not discussed, it may be associated with the potential step formed at the junction of AlGaAs-GaAs.

Some aspects of persisted photoconductivity can also be explained with the help of a frequently used model based on the presence of an energy level scheme within the band gap. Wright et al. [68] developed a model for a single crystal of CdS. The model essentially consists of a three-level energy scheme within the band gap: storage, receiving, and transport. The transport level controls the delayed and slow components of photoconductivity. The basic mechanism is relatively simple. A photon of proper energy raises an electron to a higher energy level. Ionization helps to cause deformation of the crystal near the defect center, which creates an energy state of lower energy, and the electron drops to a lower energy level. Finally, the charge carriers will be passed on to the transport state (probably impurity states), but the time required for this transition depends upon the energy difference between the receiving and transport states. Thus delay is observed in photoconductivity.

A typical suggested deformation is the transformation of the wurtzite structure to that of rock salt by spatial movement of the constituent atom. In CdS, it has been shown that a slight movement of the cadmium atom in a specific direction forms the rock salt structure. This can be achieved if the interstitial impurity atom below the sulfur plane is ionized by the radiation [68].

A modified model, based on the deformation of the lattice, has been considered by Lang and Logan [77]. The model is more suitable for polar materials such as CdS and CdTe, where there exists a strong interaction between electrons and phonons. The origin of PPC has been attributed to the drastic rearrangement of the defect configuration. It is found that the total energy of the system (electronic plus deformation) has a parabolic dependence on configurational coordinates. The energy diagram generally used for macroscopic models of lattice relaxation is given in Fig. 7.14 [77]. Curves V and C correspond to the total energy of the unoccupied defect with a delocalized electron in the valence and conduction bands, respectively. Curve D represents the energy corresponding to an occupied defect. In large relaxation approximation, the defect level is situated above the conduction band minimum (equilibrium condition), but when the energy is supplied, the local environment is distorted and the defect energy level drops as shown in Fig. 7.14. Even though the drop is small, the defect configuration coordinate is changed, and it has an important consequence. Since the optical transition takes place at the defect configuration coordinate ($\Delta Q = 0$), for this value of ΔQ the total energy for exciting electrons from the defect level to the conduction band is substantial (transition A in Fig. 7.14).

The probability of this transition, therefore, is very small; that is, it will take quite a long time for an electron from a defect state to go to the conduction band. In short, the transition through the energy state caused by deformation of the defect can give rise to the observed PPC in some semiconductor compounds, particularly polar semiconductors.

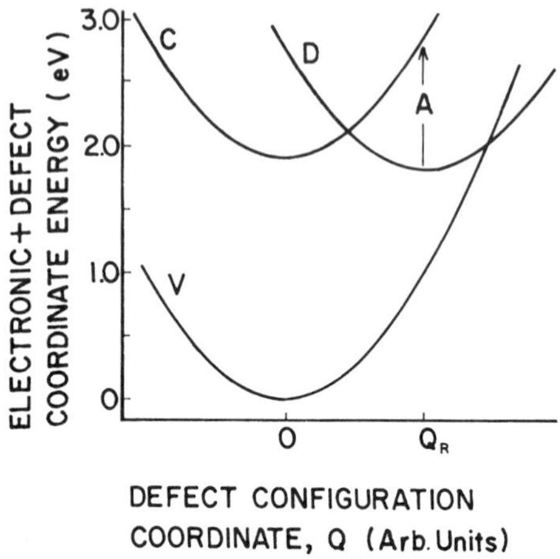

FIGURE 7.14 Variation of total energy (electronic and defect configuration) versus defect configuration coordinate. [After Ref. 77.]

Independent of the details, both models consider the deformation in the lattice and the crucial role of the energy state corresponding to it. However, the proposed lattice deformation has generally not been directly verified, and in some compounds and some structures such deformation might not even be possible.

Joshi took a completely different approach to explaining PPC [78] according to which the slow component in the photoconductivity arises as a consequence of the asymptotic stability of the differential equation system involved in the kinetics of the photoconductivity process. As mentioned in Chapter 5, there is an interplay of electrons between the conduction band and the traps. In the situation where $c_1/\gamma_1 < 1$, where c_1 is the capture cross section for electrons and γ_1 is the probability of ejection of the electron from the trap to the conduction band, the electrons remain in the trap or defect center for a considerable period of time and continue to eject electrons to the conduction band according to the ratio c_1/γ_1. This makes the photoresponse persist for a longer period of time. This is the only model that explains PPC without any kind of potential barrier in the system. The important feature of this process is that photocurrent decays slowly with nonperiodic oscillations in it.

Very often, the experimentalist observes slow, oscillatory decay response even in some commercially available photoconductivity detectors, for example, $Hg_xCd_{1-x}Te$, but such results have not been analyzed in detail.

Thus, a simple model explains the PPC effect without considering the deformation of the crystal or the extra defect states within the band gap. On the other hand, PPC is a direct and natural consequence of the asymptotic stability of nonlinear differential equations, and its presence and duration depend upon the relative values of the trap parameters.

7.7. PHOTOELECTRIC FATIGUE

Photoelectric fatigue is a rarely investigated aspect of photoconductivity. It also has a slow component of relaxation, but it differs from the previous one (PPC) in the sense that the magnitude of the sensitivity decreases considerably, sometimes by a few orders of magnitude, during the illumination process itself [70,79,80]. A typical curve of photosensitivity is shown as a function of time in Fig. 7.15.

Here the photoresponse reaches its maximum value at the start and then suddenly declines (nonexponentially); afterwards, the conductivity relaxes to its original value in various ways. It may relax slowly or instantaneously [79]; it may pass through a minimum or attain its dark current value smoothly.

Several features of relaxation curves are similar to those of PPC, but in the photoelectric fatigue the intensity of the radiation that causes fatigue is higher than the threshold value, which varies from material to material.

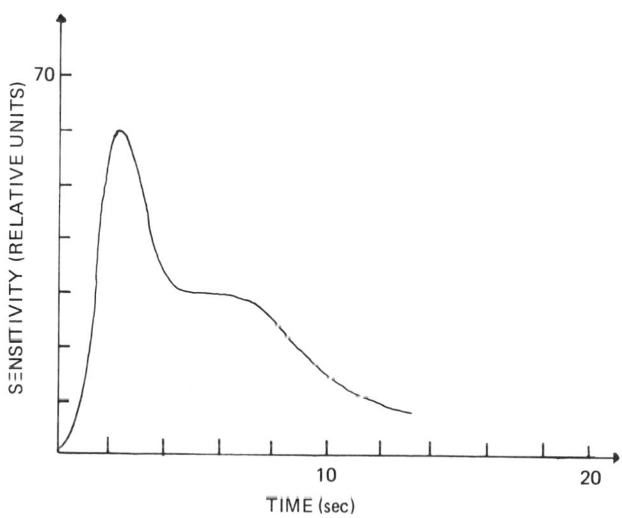

FIGURE 7.15 A demonstrative example of sensitivity variation as time advances.

Photoelectric fatigue has been observed in only a few photoconductors, including doped and undoped CdS, Cu_2O, and GaSb:S [70, 79,80]. The view put forward to explain this phenomenon is that the intense radiation creates photochemical reactions or noticeable deformations, but no satisfactory models have been developed so far.

7.8. POTENTIAL MATERIALS FOR VISIBLE AND INFRARED DETECTORS

We have seen in Chapter 4 that 3d orbitals of transition metal ions such as manganese and iron (group VIIA) maintain their symmetries even after they are introduced into II-VI compounds. Their energy states are split according to site symmetry and lie somewhere near the valence band. The transitions originating from them take part in the photoconduction process; because of this property, among many others, the applicability of these materials to photoconducting and photovoltaic devices should be extended. Further analysis for this purpose is required.

7.8.1. II-VI Compounds with Manganese—A New Family of Semimagnetic Semiconductors

In the last few years, ternary alloys of II-VI compounds with manganese have attracted great attention for several reasons. First, important properties such as the energy band gap and the effective mass can be controlled by varying the proportion of manganese in the compound. For example, in $Cd_{1-x}Mn_xTe$ and in $Cd_{1-x}Mn_xSe$ the band gaps can be varied from 1.5 eV to 2.3 eV and from 1.68 eV to 2.22 eV, respectively, at 300 K [82]. The variation of the band gap in iron-based compounds has also been reported [83]. Because of the presence of the magnetic ion Mn^{2+}, diluted magnetic semiconductors (DMSs) display properties of disordered magnetic alloys such as spin glass transition and the formation of antiferromagnetic clusters. Moreover, spin—spin interaction between localized magnetic moments and free electrons in the conduction band give rise to dramatic changes in the electronic band structure, donor—acceptor states, and other related properties. Finally, as mentioned in Chapter 4, the atomic levels corresponding to 3d orbitals maintain their characters, the crystal field splitting takes place according to the site symmetry (T_d), and d—d transitions make their appearance. The last property is of particular interest for photodetection.

Experimental investigations on photoconducting properties of DMS materials have been carried out, and some useful information about their capabilities has been reported [84,85]. In some cases such as

$Cd_{1-x}Mn_xTe$ [85], $Hg_{1-x}Mn_xTe$ [86], and $Cd_{1-x}Mn_xS$ [87], a high photoresponse and performance comparable to that of conventional photoconductors have been observed; unfortunately, these materials have not yet been examined in the device context. The reasons for this, probably, were the lack of good quality material in thin-film form and the lack of a defect-free, reliable p-n junction. The first impediment has been overcome, and very recently superlattice structures of DMS have been reported [88].

The second difficulty is also likely to be overcome in the near future, and hence a brief summary of the experimental results follows to encourage further study on electro-optical properties of these materials.

Positions of d levels of manganese atoms (or transition metal ions) create an interesting situation. In some cases manganese levels are very close to the bottom of the valence band or within the band gap and interfere with fundamental transitions. Photoconductivity studies show that this occurs in $Cd_{1-x}Mn_xS$ [87] and in $Cd_{1-x}Mn_xSe$ [89], where the 3d levels of manganese are located at 1.91 eV and 1.9 eV, respectively, below the bottom of the conduction band. Since these levels are atomic-like, their positions are independent of both the concentration of manganese and the temperature, and obviously the photoresponse is also. This immediately suggests a unique useful feature; one can control the lattice parameter of a DMS on which a desired material can be grown epitaxially without shifting the peak response of the photodetector. This is indeed a significant property for the design and construction of photoconducting devices. If 3d levels lie significantly below the top of the valence band, the response is extended toward the high-energy side and the material becomes adequate for detecting ultraviolet radiation.

Diluted magnetic semiconductor materials for photodetection have been examined, but not extensively. For example, new and spectacular effects in the transport and optical properties due to exchange interactions in DMSs have not yet been expoloited, and experimental work in this direction is required.

In a DMS, the manganese ion is substituted randomly over the cation sites. Five electrons of the half-filled 3d shell of the manganese ion are aligned and the spins have the same direction, giving rise to a magnetic moment of 5 μ_B, where μ_B is the Bohr magneton. Thus, in addition to photoconducting properties, a DMS possesses magnetic properties that II-VI semiconducting compounds lack.

The exchange interaction causes different behavior in those experiments where Landau structure is involved, that is, magneto-optical effects and magnetotransport in the quantum limit. A typical example is a giant negative magnetoresistance. It has been found experimentally [90] that in p-type $Hg_{1-x}Mn_xTe$ at low temperatures the resistivity decreases by seven orders of magnitude in only 7 tesla, and this has been attributed to the spatial growth of the acceptor wave function in the magnetic field. This effect has important consequences for the transport of charge carriers. First, the ionization energy of the acceptors is reduced with the applied magnetic field, leading to a substantial increase in the number of charge carriers. This effect is unusual and is sometimes called the boil-off effect. The spatial extension of the acceptor wave function increases the probability of its superposition with the wave function of the neighboring acceptor, and the impurity conductivity increases. Some work has been done on these aspects [91]. Even though a limited amount of experimental work has been reported, the results are encouraging and significant. These materials are ideal for developing photoelectromagnetic detectors.

For detectors in the visible range, the DMS superlattice structure seems to have a promising future, and in the past decade experimental investigations of several aspects have been reported [91–99]. Optical properties of the $Cd_{1-x}Mn_xTe/CdTe$ multiple quantum well have been examined by Nurmikko et al. [92], and some interesting features have been observed, among them magneto-optical properties and weak confinement of holes. What is worth mentioning is the order of the magneto-optical effect. Kolodziejski et al. [94] have demonstrated that the order is several hundred times more than that of conventional semiconductors of a comparable band gap. Further, it has also been observed that optical properties depend strongly upon the crystallographic orientation of the material, so this is another controllable parameter that can be used to tailor the properties of the device.

To the device designers, any help in growth technology is a great asset, and DMS materials constitute a special tool for obtaining information about the distribution of manganese ions in the lattice. This can be achieved, in principle, through specific heat measurements with and without a magnetic field. In zero magnetic field, isolated manganese ions do not contribute to specific heat, but a pair of them sufficiently close to each other do, and this can be used to evaluate the manganese distribution. A proper distribution in the desired region can be obtained by controlling the growth conditions.

These additional features are uncommon in photodetector technology. If these properties are adequately exploited, photoelectromagnetic detectors (visible as well as infrared) will have new applications, both in industry and in research, and this investigation has yet to be carried out.

7.9. PRESENT AND FUTURE

We have seen that photoconductivity is a complex phenomenon in which several processes are simultaneously involved. The situation is further complicated by the fact that many of these processes are themselves intricate and not yet fully understood. In spite of this, because of over a hundred years of intensive work in this field, the general mechanism is reasonably clear and has been summarized here. Conventional, well-accepted approaches along with modern debatable views have been presented.

In this decade, we are witnessing rapid development in quantum well and superlattice structures where entirely different principles, such as charge carriers in lower dimensions, are operating. Developments over the last couple of years have nevertheless had a tremendous impact on improving the general performance of modern photosensors. Looking at these developments and considering the demands of fiber-optical communications and the integrated electro-optics of photonic devices, we expect to see major breakthroughs in this area within a short time.

REFERENCES

1. A. K. Ramdas and S. Rodriguez, Rep. Prog. Phys. 44,1297 (1981).
2. R. A. Stradling, in New Developments in Semiconductors, P. R. Wallace, R. Harris, and M. J. Zuckermann, Eds., Noordhoff.
3. P. E. Simmonds, R. A. Stradling, J. R. Birch, and C. C. Bradley, Phys. Stat. Solidi b 64,195 (1974).
4. S. T. Pantelides, Rev. Mod. Phys. 50,797 (1978).
5. C. Jagannath, Z. W. Grabowski, and A. K. Ramdas, Phys. Rev. B 23,2082 (1981).
6. R. C. Milward, in Far Infrared Properties of Solids. S. S. Mitra and S. Nudelman, Eds., Plenum, New York, 1970, p. 1.
7. R. J. Bell, Introductory Fourier Transform Spectroscopy. Academic, New York, 1972.
8. S. S. Mitra and S. Nudelman, Eds., Far Infrared Properties of Solids. Plenum, New York, 1970.
9. J. H. Reuszer and P. Fischer, Phys. Rev. 140A,245 (1965).
10. N. V. Joshi and A. B. Vincent, Phys. Lett. 92A,476 (1982).
11. L. T. Ho and A. K. Ramdas, Phys. Rev. B 5,462 (1972).
12. J. H. Reuszer and P. Fisher, Phys. Rev. A 135,1125 (1964).
13. A. A. Kaplyanskii, Sov. Phys.-Opt. Spectr. 16,557 (1964).
14. B. Lax, L. M. Roth, and S. Zwerdling, J. Phys. Chem. Solids 8,311 (1959).
15. R. R. Hearing, Can J. Phys. 36,1161 (1958).

16. H. R. Chandrasekhar, A. K. Ramdas, and S. Rodriguez, Phys. Rev. B 14,2417 (1976).
17. S. M. Kogan and B. I. Sedunov, Sov. Phys. Solid State 8, 1898 (1967).
18. G. E. Stillman, C. M. Wolf, and J. O. Dimmock, Far infrared photoconductivity in high purity GaAs, in Semiconductors and Semimetals, Vol. 12. R. K. Willardson and A. C. Beer, Eds. Academic, New York, 1977, p. 291.
19. S. D. Seccombe and D. M. Korn, Solid State Commun. 11,1539 (1972).
20. R. J. Wanger and B. D. McCombe, Phys. State. Solidi b 64, 205 (1974).
21. N. V. Joshi, Solid State Commun. 36,509 (1980).
22. R. K. Watts, Point Defects in Crystals. Wiley, New York, 1976.
23. P. J. Kireev, Semiconductor Physics. MIR, Moscow, 1978.
24. N. V. Joshi and A. B. Vincent, Phys. Lett. 92A,476 (1982).
25. P. E. Simmonds, R. A. Stradling, J. R. Birch, and C. C. Bradley, Phys. Stat. Solidi 64,195 (1974).
26. H. Y. Fan, Proc. Int. Conf. Physics of Semiconductors, Moscow, 1968, p. 135.
27. H. J. Stocker, H. Levinstein, and C. R. Stanndard, Phys. Rev. 150,613 (1966).
28. A. L. Mears, A. R. L. Spray, and R. A. Stradling, J. Phys. C1,1412 (1968).
29. R. H. Bube, Photoconductivity of Solids, Wiley Interscience, New York, 1960.
30. V. V. Zotov and V. V. Serdyuk, Phys. Stat. Solidi 28,K31 (1968).
31. S. O. Hemila and R. H. Bube, J. Appl. Phys. 38,5258 (1967).
32. R. H. Bube and F. Cardon, J. Appl. Phys. 35,2712 (1964).
33. A. Rose, Concepts in Photoconductivity, Interscience Tracts on Physics and Astronomy, Vol. 19, Wiley Interscience, New York, 1963.
34. A. Rose, Phys. Rev. 97,322 (1955).
35. E. Yamaichi, Y. Akishige, and K. Ohi, Jpn. J. Appl. Phys. 23,867 (1984).
36. N. V. Joshi and R. Echeverria, Solid State Commun. 47,251 (1983).
37. J. L. Shay and J. H. Wernick, Ternary Chalcopyrite Semiconductors: Growth, Electronic Properties and Applications. Pergamon, Oxford, 1985.
38. M. J. Thwaites, R. D. Tomlison, and M. J. Hampshire, Inst. Phys. Conf. Ser. 35,237 (1977).
39. J. E. Jaffe and A. Zunger, Phys. Rev. B 28,5822 (1983).
40. J. E. Jaffe and A. Zunger, Phys. Rev. B 29,1882 (1984).

41. N. V. Joshi, J. Sanchez, and J. Martin, J. Phys. and Chem. of Solids 50,629 (1989).
42. R. Echeverria, A. B. Vincent, and N. V. Joshi, Solid State Commun. 52,901 (1984).
43. D. Redfield and J. P. Wittke, Proceedings of the Third International Conference on Photoconductivity. E. M. Pell, Ed. Pergamon, Oxford, 1971, p. 29.
44. V. V. Serdyuk, L. E. Stys, A. E. Turetskii, C. G. Chemeresyuk, and A. M. Shmilevich, Sov. Phys. Semicond. 18,86 (1984).
45. E. A. Davis, Solid State Electron. 9,605 (1966).
46. L. Forbes and C. Sah, Solid State Electron.14,182 (1971).
47. W. Siemens, Wied Ann. Phys. 2,521 (1977).
48. J. Stuke, Phys. Stat. Solidi 6,441 (1964).
49. R. H. Bube, Phys. Rev. 99,1105 (1955).
50. J. Dresner, J. Chem. Phys. 35,1628 (1961).
51. I. Ikovich, C. Viger, and C. Vautier, Proc. of 5th Conf. on Amorphous and Liquid Semiconductors, Taylor & Francis, London, 1974, p. 817.
52. F. Stockmann, Z. Physik. 143,348 (1955).
53. S. G. Abdullaev, V. A. Aliev, N. T. Mamedov, and M. K. Sheinkman, Sov. Phys. Semicond. 17,1141 (1983).
54. N. V. Joshi, L. Mogollon, J. Sanchez, and J. M. Martin, Solid State Commun., 65:151 (1988).
55. R. A. Hopfel, Appl. Phys. Lett. 52,801 (1988).
56. A. L. Robinson and R. H. Bube, J. Appl. Phys. 42,5280 (1971).
57. S. G. Patil, J. Phys. D 5,1692 (1972).
58. T. Yoshida, T. Oka, and M. Kitagawa, Appl. Phys. Lett. 21,1 (1972).
59. S. Radhakrishnan, M. N. Kamalasanan, and P. C. Mehendru, J. Mat. Sci. 18,1912 (1983).
60. V. T. Mak, Sov. Phys. 26,558 (1983).
61. A. Matsuda, Jpn. J. Appl. Phys. 25,L54 (1986).
62. N. V. Joshi, C. Rodriguez, and A. B. Vincent, Nuovo Cimento 2D,1906 (1983).
63. A. B. Vincent, C. Rodriguez, and N. V. Joshi, Can. J. Phys. 62:883 (1984).
64. J. Martin, R. Casanova, and N. V. Joshi, Phys. Rev. 36:9703 (1987).
65. L. Fiermans, J. Vennicks, and W. Dekeyser, Electron and Ion Spectroscopy of Solids. NATO Advanced Study Inst. Ser., Vol. 32. Plenum, New York, 1977.
66. G. Margaritondoy and J. H. Weaver, in Methods of Experimental Physics Vol. 22. R. L. Park and M. G. Lagally, Eds. Academic, New York, 1985, p. 127.
67. T. Takahashi and A. Ebina, Appl. Surface Sci. 11/12,268 (1982).

68. H. C. Wright, R. J. Downey, and J. R. Canning, J. Phys. D 1,1593 (1968).
69. A. Ya Yul, L. V. Golubev, L. V. Sharonov, and Yu V. Shmartsev, Sov. Phys. Semicond. 4,2017 (1971).
70. M. K. Sheinkman and A. Ya Shik, Sov. Phys. Semicond. 10, 128 (1976).
71. G. W. Eseler, J. A. Kafalas, A. J. Strauss, H. F. MacMillan, and R. H. Bube, Solid State Commun. 10,619 (1972).
72. B. C. Burkey, R. P. Khosla, J. R. Fischer, and D. L. Losee, J. Appl. Phys. 47,1095 (1976).
73. J. W. Farmer and D. R. Locker, J. Appl. Phys. 52,5718 (1981).
74. H. J. Queisser and D. E. Theodorou, Phys. Rev. Lett. 43,401 (1979).
75. M. R. Lorenz, B. Segall and H. M. Woodbury, Phys. Rev. 134, A751 (1964).
76. H. J. Queisser, 17th Conf. of Physics of Semiconductors.
77. D. V. Lang and R. A. Logan, Phys. Rev. Lett. 39,635 (1977).
78. N. V. Joshi, Phys. Rev. 32,1009 (1985).
79. N. B. Lukkyanchikova, A. A. Konoval, and M. K. Sheinkman, Solid State. Electron. 18,65 (1975).
80. A. I. Andirevskii and A. L. Rvachev, Photoelectric and Optical Phenomena in Semiconductors [in Russian], Izv. Akad. Nauk Ukr. SSR Kiev, 1959, p. 164.
81. N. Bottka, J. Stankiewicz, and W. Giriat, J. Appl. Phys. 52, 4189 (1981).
82. W. Giriat, Sixth Int. Conf. on Ternary and Multinary Compounds, in Crystal Growth and Characterization, vol. 10, B. R. Pamplin, N. V. Joshi, and C. Schwab, Eds., Pergamon, Elmsford, New York, 1984, p. 45.
83. N. V. Joshi and L. Mogollon, Sixth Int. Conf. on Ternary and Multinary Compounds, in Crystal Growth and Characterization, vol. 10, B. R. Pamplin, N. V. Joshi, and C. Schwab, Eds., Pergamon, Elmsford, New York, 1984, p. 65.
84. J. K. Furdyna, J. Appl. Phys. 53,7637 (1982).
85. N. V. Joshi, J. Martin, and P. Quintero, Appl. Phys. Lett. 39,79 (1981).
86. R. R. Galazka and J. Kossut, in Narrow Gap Semiconductors: Physics and Applications, Lecture Notes in Physics Ser., No. 133. Springer-Verlag, New York, 1980, p. 245.
87. J. M. Martin, A. B. Vincent, and N. V. Joshi, Proc. SPIE 395,267 (1983).
88. R. L. Gunshor, N. Otuska, M. Yamanishi, L. A. Kolodziejski, T. C. Bonsett, R. B. Bylsma, S. Datta, W. M. Becker, and J. K. Furdyna, J. Crystal Growth 72,294 (1985).
89. N. V. Joshi, L. G. Roa, and A. B. Vincent, Nuevo Cimento 2D,1880 (1983).

90. A. J. Mycielski and J. Mycielski, J. Phys. Soc. Jpn. 49,807 (1980).
91. J. K. Furdyna, J. Vac. Sci. Technol. 21,220 (1982).
92. A. V. Nurmikko, R. L. Gunshor, and L. A. Kolodziejski, IEEE J. Quantum Electron. QE22,1785 (1986).
93. J. P. Faurie, IEEE J. Quantum Electron. QE22,1656 (1986).
94. L. A. Kolodziejski, R. L. Gunshor, N. Otsuka, S. Datta, W. M. Becker, and A. V. Nurmikko, IEEE J. Quantum Electron., QE22,1666 (1986).

Appendix

Some Useful Photoconductors and Their Band Gaps

Type	Material	Band Gap (eV)
Elements	α-Sn	0.08
	Te	0.34
	Ge	0.68
	Si	1.10
	Se	2.1
Binary Compounds		
I-V	Cs_3Bi	0.5
	CsSb	0.8
	KSb	0.9
	Rb_3	1.0
	K_3Sb	1.1
	Na_3Sb	1.1
	Cs_3	1.6
I-VI	Ag_2S	0.9
	Cu_2O	2.0
I-VII	AgI	2.8
	CuBr	2.9
II-IV	Mg_2Sn	0.3
	Ca_2Pb	0.46
	Mg_2Ge	0.6
	Mg_2Si	0.7
	Ca_2Sn	0.9
	Ca_2Si	1.9

Some Useful Photoconductors and Their Band Gaps

Type	Material	Band Gap (eV)
II-V	CdSb	0.5
	ZnSb	0.5
	Cd_3As_2	0.55
	Cd_3P_2	0.55
	ZnSb	0.56
	Mg_3Sb_2	0.82
	Zn_3As_2	1.00
	Zn_3P_2	1.15
II-VI	HgTe	0.2
	HgSe	0.6
	CdTe	1.5
	CdSe	1.7
	HgS (red)	2.0
	ZnTe	2.1
	CdS	2.4
	ZnO	3.2
	BaTe	3.4
	ZnS	3.7
	BaSe	3.7
	BaS	4.0
	SrTe	4.0
	BaO	4.2
	CaTe	4.3
	SrSe	4.6
	MgTe	4.7
	SrS	4.8
	CaSe	5.0
	CaS	5.4
	MgSe	5.6
	SrO	5.8
II-VII	HgI_2	2.55
III-V	InSb	0.18
	InAs	0.33
	GaSb	0.67
	InP	1.25
	GaAs	1.4
	AlSb	1.5
	GaP	2.24
	AlAs	2.4
	GaN	3.4

Some Useful Photoconductors and Their Band Gaps

Type	Material	Band Gap (eV)
III-VI	In_2Te_3	1.0
	Tl_2S	1.2
	In_2Se_3	1.2
	InSe	1.2
	Ga_2Te_3	1.2
	GaTe	1.5
	Ga_2Se_3	1.9
	GaSe	2.0
	In_2S_3	2.0
	GaS	2.0
	Ga_2S_3	2.5
	Al_2Te_3	2.5
	Al_2Se_3	3.1
	In_2O_3	3.5
	Al_2S_3	4.1
	Ga_2O_3	4.4
	Al_2O_3	> 5.0
IV-V	SiC (cub)	2.3
	SiC (hex)	2.9
IV-VI	PbSe	0.25
	PbTe	0.31
	PbS	0.40
	SnS	1.3
	PbO	2.3
	TiO_2	3.0
	SnO_2	4.3
V-VI	Bi_2Te_3	0.15
	Sb_2Te_3	0.3
	Bi_2Se_3	0.35
	As_2Te_3	1.1
	Sb_2Se_3	1.2
	Bi_2S_3	1.3
	Sb_2S_3	1.7
	As_2Se_3	1.7
	AS_2S_3	2.5
	Bi_2O_3	3.2
	As_2O_3	4.0
	Sb_2O_3	4.2
VI-VI	MoS_2	1.2
	TeO_2	1.5

Some Useful Photoconductors and Their Band Gaps

Type	Material	Band Gap (eV)
Ternary	$CuInTe_2$	≃ 0.96
	$AgInTe_2$	≃ 1.0
I-III-VI_2	$CuInSe_2$	1.04
	$AgGaTe_2$	≃ 1.2
	$CuGaTe_2$	1.23
	$AgInSe_2$	1.24
	$CuInS_2$	1.53
	$CuGaSe_2$	1.68
	$AgGaSe_2$	1.83
	$AgInS_2$	≃ 1.87
	$CuAlTe_2$	≃ 2.06
	$AgAlTe_2$	2.27
	$CuGaS_2$	2.43
	$AgAlSe_2$	2.55
	$AgGaS_2$	≃ 2.6
	$CuAlSe_2$	2.67
	$AgGaS_2$	2.73
	$AgAlS_2$	3.13
	$CuAlS_2$	3.49
II-III_2-VI_4	$CdIn_2Te_4$	0.94
	$CdIn_2Se_4$	1.72
	$ZnIn_2Se_4$	1.82
	$HgGa_2Se_4$	1.95
	$HgIn_2S_4$	2.00
	$CdIn_2S_4$	2.3
	$CdGa_2Se_4$	2.43
	$ZnIn_2S_4$	2.6
	$HgGa_2S_4$	2.84
	$CdGa_2S_4$	3.44
II-IV-V_2	$CdSnAS_2$	0.26
	$CdGeAS_2$	0.57
	$ZnSnAS_2$	0.73
	$ZnGeAS_2$	1.15
	$CdSnP_2$	1.17
	$CdSiAS_2$	1.55
	$ZnSnP_2$	1.66
	$CdGeP_2$	1.72
	$ZnSiAS_2$	2.12
	$CdSiP_2$	2.45
	$ZnGeP_2$	2.34
	$ZnSiP_2$	2.96

Some Useful Photoconductors and Their Band Gaps

Type	Material	Band Gap (eV)
Quaternary	$CuInSnSe_4$	0.71
	$AgInSnSe_4$	0.94
	$CuInGeSe_4$	1.26
	$CuGaSnSe_4$	1.42
	$AgInGeSe_4$	1.58
	$AgGaSnSe_4$	1.70
	$AgGaGeSe_4$	1.85
	$AgAlSnSe_4$	1.85
	$CuGaGeSe_4$	1.87
	$CuAlSnSe_4$	1.90
	$AgAlGeSe_4$	2.02
	$CuAlGeSe_4$	2.25

List of Symbols

a_0, a_1, a_2	Constants involved in the solution of differential equations
A	Area
b_0, b_1, b_2	Constants
c_0, c_1, c_2	Constants
c	Speed of light
C	Constant
$\underline{C}_n$	Capture cross section for electrons
$\underline{C}_p$	Capture cross section for holes
D	Detectivity
D*	Specific detectivity
D_e	Diffusion coefficient for electrons
D_p	Diffusion coefficient for holes
E	Electric field
E_t	Energy of the traps
E_{s-o}	Energy corresponding to spin-orbit interaction
E_g	Band-gap energy
E_A	Ionization energy of acceptors
E_D	Ionization energy of donors
E_{Ext}	Ionization energy of excitons

E_{hh}	Energy of heavy-hole valence band
E_{lh}	Energy corresponding to light-hole valence band
$E_{\underline{k}}$	Energy of conduction band at wave vector $\underline{k}$
$E_{v\underline{k}}$	Energy of valence band at wave vector $\underline{k}$
E_{phon}	Energy of phonon
e	Absolute value of the electronic charge
Δf	Band width
f	Fraction of certain quantity
g_h	Generation rate of electrons due to incident radiation
g_n	Generation rate of holes
$\underline{g}$	Density of states
G	Conductance
f_{ij}	Oscillator strength
$f(E_f)$	Fermi energy
F	Figure of noise
h	Planck's constant
$\hbar$	Dirac's constant, $\hbar = h/2\pi$
$\underline{I}$, $\underline{I}_0$	Intensity of radiation
I	Current density
I_d	Dark current density
I_n	Current due to electrons
I_p	Current due to holes
I_{pho}	Photocurrent
I_s	Signal current (independent of the origin)
$I_{d,noise}$	Dark current noise
k	Boltzmann constant
K	Absolute temperature scale
$\underline{k}$	Wave vector
L_e	Diffusion length for electrons
L_h	Diffusion length for holes
J	Electron flux
m_h	Mass of hole

m_e	Mass of electron
m^*	Effective mass of electron in the conduction band
m_r^*	Effective reduced mass
M_{ij}	Matrix corresponding to transition between states i and j
n	Number of electrons per cubic centimeter at equilibrium
$n(t)$	Number of electrons per cubic centimeter at time t
n_c	Number of occupied traps per cubic centimeter
$\underline{n}(t)$	Excess of electrons at time t
$\underline{n}_0$	Steady-state electrons in a perturbed condition
N	Number or density of countable entities
N_{pho}	Photon density
N_t	Total number of traps per cubic centimeter
N_{cm}	Effective density of free electron states in the conduction band
$\hat{O}$	Operator corresponding to electronic transitions
p	Density of holes in the valence band
p	Orbital corresponding to angular momentum l = 1
$\underline{p}$	Momentum
$\underline{p}$	Excess of holes
P	Power of incident radiation
P_k	Transition probability at k
P_n	Noise power
$\underline{P}$	Crystal momentum
r	Rate of production of photons
R_{auto}	Autocorrelation of a function
R_{en}	Rate of reemission of electrons to the conduction band
R_{BM}	Bimolecular recombination constant
R	Resistance in general
R_L	Load resistance
R_d	Resistance in dark
R_{illu}	Resistance under illumination
R_n	Rate of recombination of electrons

R_h	Rate of recombination of holes
$\underline{R}(S)$	Signal response
$R(\lambda)$	Reflectivity
R	Rydberg constant
S	Signal
T	Absolute temperature
$\langle v \rangle$	Average velocity of the electrons
$\langle \ \rangle$	Average value
V	Volume element
W	Width
$\underline{\alpha}$	Impact ionization parameter for electrons and holes
α	Absorption coefficient
$\boldsymbol{\alpha}$	Type of traps
α_1, α_2	Reciprocal time constant τ_1^{-1}, τ_2^{-1}, etc.
$\underline{\beta}$	Impact ionization parameter for holes
$\boldsymbol{\beta}$	Type of traps
γ	Responsivity
γ_1	Probability of release of a trapped electron by thermal agitation
δ_0	Nonradiative recombination by capturing a hole from the valence band
ε	Dielectric constant
ε_0	Permittivity of free space
η	Quantum yield
μ	Frequency of radiation
μ_n	Mobility of the electrons
μ_p	Mobility of the holes
ρ	Excess of space charge
σ	Conductivity
$\underline{\sigma}$	Stefan-Boltzmann constant
$\underline{\hat{\sigma}}$	Pauli spin operator
τ_{bulk}	Bulk lifetime constant
τ_{coll}	Average time between two successive collisions

τ_{decay}	Decay time of a photoconductor
τ_{diel}	Dielectric time constant
$\tau_{lifetime}$	Average time spent by the electron or hole in the conduction or valence band, respectively
τ_{maj}	Majority carrier lifetime
$\tau_{minority}$	Minority carrier lifetime
τ_{rel}	Relaxation time constant
τ_{rise}	Rise time of a photoconductor
$\tau_{surface}$	Surface time constant
$\tau_{transit}$	Time required to reach the electrode
ϕ	Radiation flux
$\underline{\phi}$	Electrostatic potential
ϕm	Vacuum work function of a metal
$\phi m\text{-}S$	Metal-semiconductor work function difference-barrier height
ϕS	Surface potential
χ	Parameter to describe the extension of wave function
ψ	Wave function
ω	Angular frequency

Index